SAGE was founded in 1965 by Sara Miller McCune to support the dissemination of usable knowledge by publishing innovative and high-quality research and teaching content. Today, we publish more than 750 journals, including those of more than 300 learned societies, more than 800 new books per year, and a growing range of library products including archives, data, case studies, reports, conference highlights, and video. SAGE remains majority-owned by our founder, and after Sara's lifetime will become owned by a charitable trust that secures our continued independence.

Los Angeles | London | Washington DC | New Delhi | Singapore | Boston

A Comparative Study on the Role of Universities in Transformation of Knowledge and Skills in Rural Areas

Thank you for choosing a SAGE product! If you have any comment, observation or feedback, I would like to personally hear from you. Please write to me at contactceo@sagepub.in

—Vivek Mehra, Managing Director and CEO,
SAGE Publications India Pvt Ltd, New Delhi

Bulk Sales

SAGE India offers special discounts for purchase of books in bulk. We also make available special imprints and excerpts from our books on demand.

For orders and enquiries, write to us at

Marketing Department
SAGE Publications India Pvt Ltd
B1/I-1, Mohan Cooperative Industrial Area
Mathura Road, Post Bag 7
New Delhi 110044, India
E-mail us at marketing@sagepub.in

Get to know more about SAGE, be invited to SAGE events, get on our mailing list. Write today to marketing@sagepub.in

This book is also available as an e-book.

━━━━━━━━❦❧━━━━━━━━

A Comparative Study on the Role of Universities in Transformation of Knowledge and Skills in Rural Areas

Li Wang

United Nations
Educational, Scientific and
Cultural Organization

International Research
and Training Centre
for Rural Education

⑤SAGE www.sagepublications.com

<small>Los Angeles • London • New Delhi • Singapore • Washington DC • Boston</small>

First published in 2015 by

 SAGE Publications India Pvt Ltd
B1/I-1 Mohan Cooperative Industrial Area
Mathura Road, New Delhi 110 044, India
www.sagepub.in

SAGE Publications Inc
2455 Teller Road
Thousand Oaks, California 91320, USA

SAGE Publications Ltd
1 Oliver's Yard, 55 City Road
London EC1Y 1SP, United Kingdom

SAGE Publications Asia-Pacific Pte Ltd
3 Church Street
#10-04 Samsung Hub
Singapore 049483

Published by Vivek Mehra for SAGE Publications India Pvt Ltd, typeset in 10/13 Berkeley by RECTO Graphics, Delhi and printed at Chaman Enterprises, New Delhi.

Library of Congress Cataloging-in-Publication Data Available

ISBN: 978-93-515-0119-0 (HB)

The SAGE Team: Rudra Narayan, Vandana Gupta, Neena Ganjoo and Anju Saxena

Contents

Chapter 8
The Common Factors in the Cases   203

List of Figures

List of Tables

List of Abbreviations

AAA	Agriculture Advancing Australia
AAACE	Australia Association of Adult and Community Education
ABS	Australian Bureau of Statistics
AFFA	Department of Agriculture, Fisheries and Forestry, Australia
AMRI	AUH Mountain Research Institute
ANTA	Australian National Training Authority
APPEAL	Asia-Pacific Programme of Education for All
AUH	Agricultural University of Hebei, China
BIITE	Batchelor Institute of Indigenous Tertiary Education
CAEFS	Certificate in Access to Employment and Further Study
CCP	Chinese Communist Party
CDU	Charles Darwin University, Australia
CHIC	Certificate in Horticulture for Indigenous Community
CINCRM	Centre for Indigenous Natural and Cultural Resource Management
CRC	Cooperative Research Centre
CSIRO	Commonwealth Scientific and Industrial Research Organization
CTLDEC	Centre for Teaching and Learning in Diverse Educational Contexts
CUTSD	Committee for University Teaching and Staff Development
DBIRD	Department of Business, Industries and Resource Development
DEET	Department of Employment, Education and Training
DETYA	Department of Education, Training and Youth Affairs
DIT	Darwin Institute of Technology
DK-CRC	Desert Knowledge Cooperative Research Centre
EHS	Faculty of Education, Health and Science, CDU

FI	Field Visit and Investigation
ESOs	Essential Services Operation
GCs	Graduate Catalysts
GDP	Gross Domestic Product
GSP	Gross State Product
IESP	Indigenous Education Strategic Plan
IDL	Interactive Distance Learning
IGA	Income Generating Activities
IGP	Income Generating Project
In	Interview
INRULED	International Research and Training Centre for Rural Education
ITAFE	Institute of TAFE
NGO	Non Government Organization
NT	Northern Territory, Australia
NTCE	Northern Territory Certificate for Education
NTCOSS	Northern Territory Council of Social Service
NTRC	Northern Territory Rural College
NTU	Northern Territory University
PPE	Popular Primary Education
QCOSS	Queensland Council of Social Service
R & DC	Research and Development Corporation
SAP	Student Action Programme
SEAMEO	Southeast Asian Ministers of Education Organization
SEAMEO INNOTECH	Southeast Asian Ministers of Education Organization-Regional Centre on Educational Innovation and Technology, Philippines
SS	Survey Schedule
TAFE	Technical and Further Education
TAFETPS	Technical and Further Education Triennial Planning Submission
TERC	Tropical Ecosystems Research Centre
TMM	Taihang Mountain Model
UNDP	United Nations Development Programme
UNESCO	United Nations Educational Scientific and Cultural Organization
UNESCO PROAP	UNESCO Principal Office in Asian and the Pacific
VAS	Village Adoption Scheme
VET	Vocational Education and Training
VTC	Vocational Training Centre
WAN	Wide Area Network
WEFTSU	Weighted Equivalent Full-Time Student Unit

Foreword

Just under half of the world's population are located in rural areas. In developing countries rural populations are the majority, and many of the world's poor are located in rural areas. Despite the rapid urbanization of the world, rural environments are significantly important for everyone. Currently standing at just over 7 billion people, the world's population continues to grow, and this is placing substantial pressure on a range of resources essential for human life. Issues of adequate water supplies to sustain life are prominent, particularly under threats presented by climate change. Also, food security for the billions is looming as a major issue. All people have basic food needs, but with increasing standards of living, a growing middle class of relatively affluent people is demanding higher standards of different kinds of foods. This is placing substantial pressure upon agricultural enterprises, predominantly in rural locations.

This book is an important investigation of the roles that a university plays in sustainable rural development. Universities have long been an important part of broader society, at the cutting edge of innovation and development of new ideas and practices, in sustaining and reproducing culture and social values, in the communication of ideas and information, and in challenging and critiquing society. Universities that address aspects of agricultural science and practice have an important role to play in the transformation of rural communities through the provision of research outcomes and teaching. Additionally, some universities engage in extension education activity, working closely with and assisting farmers in the application of new ideas and practices.

In investigating the role of the university in the transformation of knowledge and skills in rural areas, Professor Wang has taken an interesting approach. He has engaged in a comparative study of two universities, located in rural and regional areas of two distinctive countries.

The Agriculture University of Hebei and Charles Darwin University are quite different institutions, and yet some common threads in terms of location, focus, and community service indicate similarities. Agriculture University of Hebei is located in Hebei Province, China. It has a major role in the provision of agricultural education and extension services for the province and beyond. In particular, the nature of its programs and research are focused on rural development and rural transformation. Agricultural activity, however, in Hebei is quite often different to agricultural activity in the Northern Territory of Australia, where Charles Darwin University is located. Agriculture in China is still dominated by small holding enterprise and activity, and while there has been substantial development in agricultural practices and standards of living of Chinese farmers, they are still essentially peasant farmers working on smallholder blocks. This contrasts with agricultural activity in the Northern Territory of Australia. There, agricultural enterprise tends to be large industrialized farming practices, which are highly mechanized. Other differences between these two locations relate to environment and climatology. The Northern Territory environment ranges from tropical through to hot, dry desert conditions, whereas Hebei experiences cold winters and warm summers. Both locations have a similar struggle with water resources, in that there is not enough, and agricultural activity has a significant impact on water reserves, particularly groundwater.

However, despite the differences in the types of running programs in environment and climatology, and in agricultural practices, between these two institutions, Professor Wang has identified substantial similarities of beneficial practice.

The book poses the question, 'What is the role of universities to bring about knowledge transformation in rural communities?' It identifies a number of aspects where such knowledge transformation occurs. Universities themselves go through processes of knowledge transformation as they engage in research and teaching, which not only impacts on the communities they serve but also on the universities themselves, and this outcome is important. If a university is to be effective in any community it also needs to go through phases of knowledge transformation. Indeed, universities themselves need to be continually transformed.

Professor Wang points out that it is critically important if a university is to be effective for it to establish real links with the lives of the people whom it serves. When universities are actively engaged in their communities, both servicing and assisting communities, and also learning from communities and working collaboratively with people, they are more likely to be effective. Again, such recognition is surely not limited to universities located in rural contexts.

Zhang Xinsheng
President of China Education Association for International Exchange
President of International Union for Conservancy of Nature (IUCN)
Chairman of the Executive Board of UNESCO (2005–2007)
Vice Ministry of Education, China (2001–2009)
Chairman of the Chinese National Commission for UNESCO (2001–2009)
Chairman of UNESCO International Research and Training Centre for Rural Education (INRULED) Board

Acknowledgments

I would like to thank and acknowledge the following contributors to the book: *A Comparative Study on the Role of Universities in Transformation of Knowledge and Skills in Rural Areas*.

Dr Darol Cavanagh, former Associate Professor, Charles Darwin University, Australia, for his valuable comments and suggestions, overall guidance, and support; Dr Greg Shaw, Associate Professor, Charles Darwin University, Australia, for his valuable comments and suggestions.

The book benefits from the inputs of the staff of the UNESCO International Research and Training Centre for Rural Education (INRULED), in particular: Miss Zhang Dian, Miss Zhang Wen, Mr Chandu Lal Chandrakar, Mr Francesco Valente, and Mr Ren Chao.

Thanks is also given to the SAGE Publications.

Finally, I wish to acknowledge all people who have helped or given me any assistance during the study and book writing, as well as spending time with me for interviews, surveys, and field visits both in China and in Australia.

1

Education for Rural Development

The research for this book was carried out over a period of five years using a comparative research methodology. The methodology requires that the researcher be familiar with the political, economic, social, demographic, historical, geographical, and cultural regimes of the countries under study. The preference is that the researcher lives in the country being researched over a sustained period. During my research, I managed to become familiar with the tenets of the methodology, filling in any gaps with interviews and survey of significant others, in both countries being researched. The countries are China and Australia. The area has been reduced to two cultural institutions, one in each country, the AUH, China, and the CDU in the Northern Territory (NT), Australia, and their impact on rural transformation.

I have decided to include the exact references to the survey data so that readers can get a feel for the responses directly from the survey instrument that I used. I will code the responses as SS so that readers can cross-check the survey responses to the questions and should they wish to, develop follow-up surveys to extend the research.

Introduction

This chapter will be developed to introduce the issues of the role of universities in transformation of knowledge and skills in rural areas within their broad mission of teaching, research, and extension work, and to examine closely how education and training, knowledge, and skills have been seen as a key instrument for shaping and fulfilling the goal of rural transformation.

At present, 3 billion or 60 percent of the people in developing countries and more than half the people in the world live in rural areas. They live in isolated and often inhospitable areas and gain part or almost all of their livelihoods from agriculture with little access to modern education, technology, and resources to improve their ability and life (International Research and Training Centre for Rural Education [INRULED], 2001, p. 1). Even with the economic and social progress in terms of income-generation and promotion of people's living standard in rural areas and "although social indicators on infant mortality, life expectancy, nutrition, and education show remarkable improvements, the proportion of people living in conditions of abject poverty is still scandalously high at about one-third of global population" (Ordonez, Kasaju, and Seshadri, 1998, p. 2). The world, therefore, should give a high priority and special attention to rural people, rural education, and rural development.

The problems facing rural education can be identified as nonattendance in school, early and heavy dropout of students, adult illiteracy, technological illiteracy, gender inequality in education, concentration of poverty in villages, lack of modern information in agricultural techniques and skills, urban–rural disparity in educational investments, and in the quality of teaching and learning (INRULED, 2001, p. XI).

"Enhancing peoples' capacities and expanding their choices in life though education will be a vital component in rural economic development" (INRULED, 2001, p. XI). For this issue, it should be mentioned that rural education could not solve all of the problems, but "Education and training are two of the most powerful weapons in the fight against rural poverty and for development" (Atchoarena and Gasperini, 2003, p. 29), and also rural education is at least one critical factor for promoting people's life in rural areas.

> It is essential to turn the continuing and inevitable transition of rural areas, often with adverse consequences on the rural economy, environment and

people's life, into an active and positive process for the development of rural communities. Educational programmes have to become a vital part of this development through committed partnerships of the government, educational institutions, communities, businesses, and civil society as a whole. (INRULED, 2001, p. XI)

The challenge of education to serve poverty alleviation and rural development must become one of the main issues of education for all (EFA) (a UNESCO Policy agenda). Not taking up this challenge is to imperil the total education-for-all effort. Some theories and practices have often been constructed, such as Robert Chambers and his book, *Rural Development: Putting the Last First, Discussed Unperceived Rural Poverty* (Chambers, 1983); Basic Education for Empowerment of the Poor (for instance, Ordonez et al., 1998); and Education for Rural Development (such as Atchoarena and Gasperini, 2003). My work does not produce a grand theory, however, it, by focusing on the links between education, especially higher education and rural development, seeks to underscore the forces of change rural communities face and how education, by equipping people with appropriate knowledge, skills, and information, can empower rural population's ability, expand their choices, and enable them to exercise these choices.

For a long time, rural development has been a universal concern in many countries, especially developing countries. Rural education has been considered as a powerful vehicle to improve the advancement of rural economy and society. Therefore, rural education should not only meet the present and short-term needs of rural areas, but also undertake the necessary work to prepare high-quality rural population for upgrading rural life and economy. For this reason, rural education and rural development have been given the strategic priority by many countries for the country's prosperity and stability, particularly in developing countries.

The Roles of Universities for Rural Development

Universities have a potential role for rural development, particularly through the dissemination of new technologies for income generation and of supplying expertise for the promotion of quality in rural life.

Experience shows that universities have a strong technical expertise that enables them to become a major vehicle in promoting development in rural areas. Meanwhile, universities and institutes are becoming the main forces for converting science and technology into productivity in agricultural production. Since the teaching targets and research work of the universities are closely related to the real production demand in the rural area, all kinds of introduction of new technology, high technology, and training programs are welcomed by the farmers and agriculture-related students. It is only wise to encourage more participation of many of them in order to obtain massive effects in terms of development of rural areas especially in developing countries.

For most of the higher educational institutions, there are two major roles and functions, namely, education and research. For some universities, a third function, extension services, is a very important issue, especially agricultural universities in developing countries. "The universities cannot sit aloof in the ivory tower separated from the actual environment" (Noguchi, 1998, p. 42). Universities are more and more expected to provide useful services to the society, particularly in developing countries.

To quote Mr Jacques Delors, Chairman of UNESCO International Commission on Education for the 21st century, in his report of "Learning: The Treasure Within":

> Nowhere is the universities' responsibility for the development of society as a whole more acute than in developing countries, where research done in institutions of higher learning plays a pivotal role in providing the basis for development programmes, policy formulation and the training of middle- and high-level human resources. The importance of local and national institutions in raising the developmental levels of their countries cannot be overemphasised.

Higher education is the dynamic power needed by any country for economic development and social progress, because universities are regarded as a cradle of nurturing high-quality personnel. In today's world, the national population quality is becoming one of the fundamental reasons which influences a nation's development potential and differentiates the socioeconomic development level from different areas in the country and among the different countries. Taking this into account, many developing countries have attached great importance to rural education, including: the higher learning of agricultural technology, the universalization of the scientific knowledge in the rural societies, the promotion of the quality of rural population, the empowerment of farmers applying science, and technology in order to change the backward situations of rural development.

There are many experiences and cases that can be examined, both successful and unsuccessful, that have been gathered by some universities that are already engaged in rural development. Therefore, it is worthwhile, both practically and theoretically, to conduct a study on how these universities have implemented their development strategies in order to learn from them and be able to draw theoretical implications that are common and conductive for general practices.

In this book, the AUH, China, and the CDU, Australia, have been selected as focal institutions to carry out a comparative study. This selection enables the study of one university in the Northern hemisphere and one in the Southern that focus on the issues under discussion.

Agricultural University of Hebei (AUH), China

The AUH has been selected as a case for in-depth analysis, interpretation, description, and discussion. AUH has been providing extension services to rural farmers since the late 1970s. University professors, on the basis of their scientific research and laboratory experiments, have been providing farmers with practical training, advice, and suggestions on improving crop production, developing skills in upgrading animal husbandry, planting high-yield fruit trees, etc. (Investigation 1, 2001). The AUH, China, is a provincial and comprehensive agricultural university, with comprehensive faculties and expertise in both social and natural sciences and technologies (*AUH Information Handbook*, 2001).

AUH, established in 1902, is one of the oldest agricultural institutions of higher learning in China and one of the key universities in Hebei Province with a concentration on agronomy and forestry, covering seven fields of agriculture, industry, basic sciences, economics, management, literature, and law. Located in Baoding City, the campus is only 138 km away from Beijing. There are 25 colleges in the university with 51 undergraduate programs, 24 masters, and 4 doctoral Programs. The total enrolments of students reached 21,000, with 738 master and 35 doctoral candidates.

The university consists of three campuses, one farmland, and one forestland. The main campus covers an area of 61.8 ha with a total enrolment of over 17,000. Apart from the main campus, there are two other campuses: one is located in Qinhuangdao city which is 350 km away from the main campus, with a total enrolment of 2,000, and another in Dingzhou city, 50 km away with a total enrolment of 2,000. AUH has a 300-ha farmland located only 4 km away from the main campus, a 2,700-ha forestland located 100 km away in Ixia County, Hebei Province, and 39 experiment bases of teaching-research-production dot the different regions of Hebei Province (*AUH Information Handbook*, 2001).

Agricultural universities in China compared to nonagricultural-focused universities assume a greater potential for agricultural extension education, poverty alleviation, as well as for the sustainable development of agriculture while protecting the ecological environment than do other specialist universities.

Hebei is one of the most populated and agricultural provinces in China with a vast territory, and a great disparity in soil texture and climate from north to south among the mountain areas and the plains. The educational provision levels for rural population are usually low. They often result in agricultural development in the different parts of Hebei Province varying greatly. The following map locates the area within China (Figure 1.1).

AUH has a long-time involvement with rural education and development. From the late 1970s, it has been actively participating in a comprehensive development program in Taihang Mountain areas, where there were still a lot of poor farmers and AUH has made a great contribution in education, training, and skills development for the sustainable development in this region. Apart from this program, AUH has also set up a broad network in rural and mountainous areas of Hebei and aimed to build dozens of technical bases using educational interventions to develop an alternative model of university's education, training, research, and extension for rural development which empowered farmers and served the rural communities. Besides, those technical bases can also serve as resource centers for villages to improve the quality of life within rural contexts. This initiative serves a dual purpose. It provides actual farming experiences for both teachers and students from the technical bases. These experiences contribute to the development of teaching materials, and provide a constant source of input for research. The university (AUH) thus promotes rural development and as well it has

Figure 1.1
The Map of Hebei Province

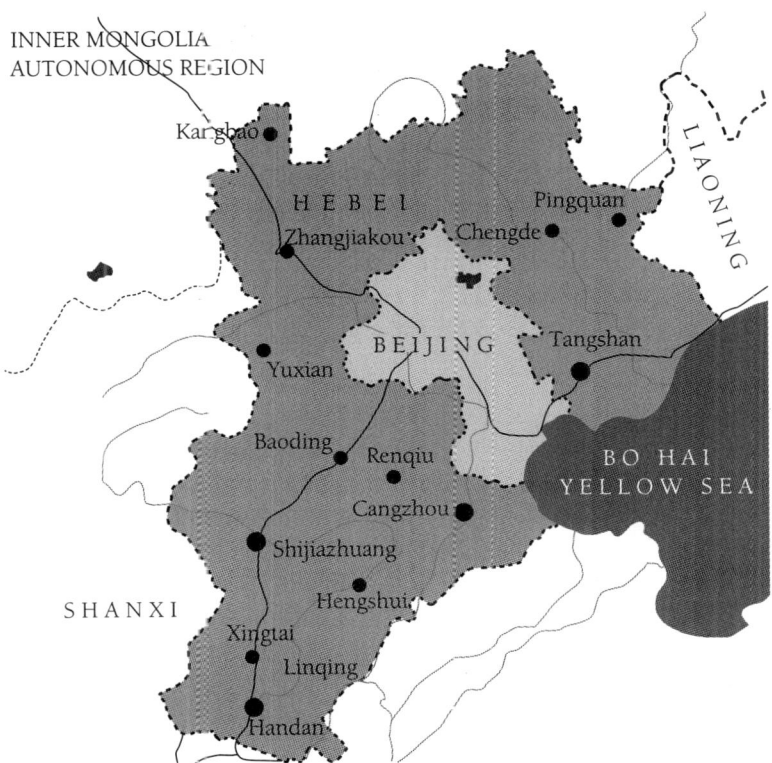

Source: http://www.map-of-china.com/hebei-s-ow.html (accessed on July 8, 2009).

enhanced its own standing and developed and extended the university capability.

It is worthwhile, therefore, to document these techniques and methods as case studies to show how the university has promoted rural development and at the same time show how the university has grown as it performs an active role in helping the rural masses improve their farm productivity and living standard. Therefore, it has earned the reputation "The Taihang Mountain Road" entitled by the Chinese government. The features of the "The Taihang Mountain Road" are: to integrate agriculture, science, and education in the rural area; to facilitate the transformation model of economic increase in the countryside; to develop the economy by means of the advancement of science and technology; and the enhancement of quality of labor. This model of rural activities is now nationally recognized in China as "Taihang Mountain Model" (Zhou, Li, and Zhang, 1990).

Since 1994, AUH trials in rural areas have been constantly monitored by external evaluators, both domestic and international, through AUH seminars and workshops. Numerous studies have also been carried out, which have confirmed that the AUH approach is a highly innovative model suitable for developing countries. A systematic study should therefore be undertaken to see if such a practice should be extended.

The Southern University in This Study Is CDU, Australia

CDU has long term been involved with education/training, research, and services for rural, remote, and Indigenous communities and made a contribution for their development.

Unlike China, Australia does not have a national education system, education being the responsibility of six separate states (New South Wales, Queensland, South Australia, Tasmania, Victoria, and Western Australia) and two territories (The Australian Capital Territory and the NT) (Anderson, 1991). Universities are established by states and are autonomous institutions; however, their funding is mostly provided by the Australian Commonwealth Government. CDU was established by an Act of the NT Parliament which was ratified by the Australian Parliament (In 19, 2004) (The actual interview can be found in Appendix 3). The CDU Ethics Committee approval was obtained for the approach, where the permission was not given, their names have been deleted.

CDU is located at Darwin, NT, Australia. The history of events leading to the foundation of the CDU (at that time, it was called Northern Territory University) begins soon after the end of the Second World War with the formation first of Darwin Adult Education Centre and then of Darwin Community College which became the Darwin Institute of Technology (DIT) at the end of 1984. By then a determined NT government had launched a long campaign for a university which came to fruition first with the opening of the University College in 1987 and then with its amalgamation with DIT to form the university (at that time, it was called Northern Territory University [NTU]) in 1989 (Berzins and Loveday, 1999). In the beginning of 2004, in order to enlarge the higher education sector to cover more territory, Alice Springs' Centralian College joined with NTU to become CDU—the only Australian University to offer the full spectrum of education options from senior secondary through to Vocational Education Training (VET) (previously known as TAFE), undergraduate, and postgraduate degree programs (http://www.cdu.edu.au/visiting/abouthistory.html, accessed on May 10, 2010).

This university and the former institute have been engaged in rural communities and Aboriginal Indigenous education for a long time. The identification of fruitful experiences and data collection could be used for a comparative study.

The NT is an autonomous territory of Australia, and one of two mainland territories of Australia, the other is the Australian Capital Territory (Shaw, 1999, p. 4). It accounts for 17.5 percent of Australia's landmass with just 1 percent of the country's population, which is 0.1 person/km² and has the lowest population density of any state (ABS [Australian Bureau of Statistics] No. 5220.0). Of this population 2 out of 3 are in the two major towns; Darwin, the capital, and Alice Springs, over 1,500 km down to the south. Approximately 25 percent of the population of the Territory are Aboriginal people (Shaw, 1999).

The NT has a wide range of institutions of post-compulsory education providing school education and training. CDU is a dual sector of educational provider that provides a range of higher education, undergraduate and postgraduate programs in areas of Arts, Fine Arts, Education, Business Hospitality and Tourism, Engineering, Information Technology, Sciences, and Law, as well as Technical and Further Education (TAFE) programs. Batchelor Institute of Indigenous Tertiary Education (BIITE), a college specifically for the educational needs of Indigenous people provides a range of higher education, diploma, and associate diploma programs as well as a range of TAFE programs. Also, there are ranges of private educational providers through church and community organizations, as well as larger companies such as Nabalco and Gemco who offer in-house training of their staff and indigenous (Shaw, 1999, p. 6). In 2011, the Batchelor Institute came under the wing of CDU and entered a collaborative partnership with CDU.

Even though most of those educational providers deliver the programs in rural, remote, and Indigenous communities, CDU as the only higher education institution in NT plays an important role for education, training, research, consultancy, and extension in rural NT. For example, the Northern Territory Rural College (NTRC) is a college belonging to

Faculty of Law, Education, Business and Arts, CDU. It primarily provides training for VET and delivers TAFE programs. It is located 16 km north of Katherine and 300 km south of Darwin. Katherine is the center of a rapidly developing mining, agricultural, horticultural, and tourist region. The college has excellent facilities located on its 4,000-ha main campus and also has its own 700-km² cattle station.

The college is currently expanding its mode of delivery to incorporate both regional and remote delivery into other NT centers to include Alice Springs and Tennant Creek and the Far West Kimberley region of Western Australia. External study and remote training packages and training workshops are constantly expanding to include delivery to cover northern Australia from Broome and Kununurra in the west and from Townsville to Cape York, in North Queensland.

The college has provided the training areas of agriculture, aquaculture, horticulture, beef cattle production, farm chemical safety and application and other chemical user program, childcare, horsemanship skills, lands, parks and wildlife management, station maintenance skill, vehicle maintenance and operation, welding, computers, new apprenticeships/traineeships, pest management, etc.

The NTRC has excellent teaching facilities with well-equipped workshops, cattle-handling facilities, stables, modern air-conditioned classrooms, and a library/computing facility located on its 4,000-ha main campus. The college also has its own 700-km² cattle station situated 90 km south of Katherine at Mataranka. The NTRC has full residential facilities for those who attend the NTRC and the CDU Regional Centre in Katherine. Students are provided with individual rooms, all meals, and a full laundry service. A full-time residential supervisor is responsible for the personal welfare of students. Students with medical problems are referred immediately to the local hospital or the private medical clinic. College staff is on duty after hours to provide guidance and supervision to all students.

The tropical agriculture, horticulture and aquaculture sections can provide training packages customized to suit the needs of your horticultural or aquaculture industry. Customized training can upgrade the skills of the current staff to meet the immediate demands of a customer's workplace. Flexibility in delivery is the key. Training can be offered in the workplace at times that best suit customer's enterprise (NTRC pamphlet).

The following maps locate the area within Australia (Figures 1.2 and 1.3).

Figure 1.2
Map of Australia and Location of NT

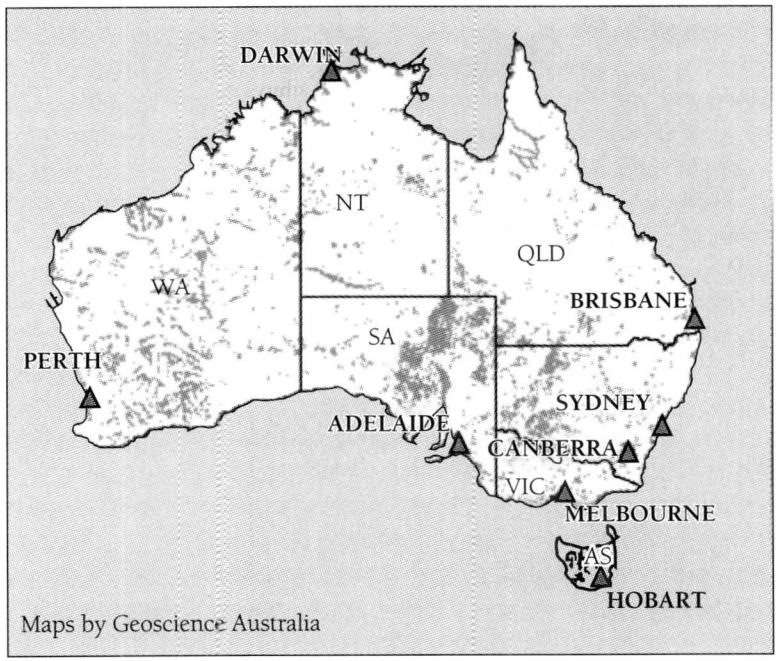

Maps by Geoscience Australia

Source: http://www.expedia.com/pub/agent.dll (accessed on July 8, 2009).

Other Universities in Developed and Developing Countries for Rural Development

The history of the developed countries and the roles of the university for rural development show that the agricultural universities have been taking a key role in serving rural communities. About 140 years ago, for example, many universities in America were agriculture-oriented universities. In the initial stage, most of them were land-grant universities, to name a few of them, Maryland State University, Cornell University, and so on. Specifically, the law required each land grant university to offer programs for working farmers and homemakers in agricultural and relevant sciences. Those universities continue the strong concern for agricultural research work through agricultural extension for the development of the national economy in America (http://encyclopedia.thefreedictionary.com/land-grant%20university, accessed on July 16, 2009).

Figure 1.3
Map of NT

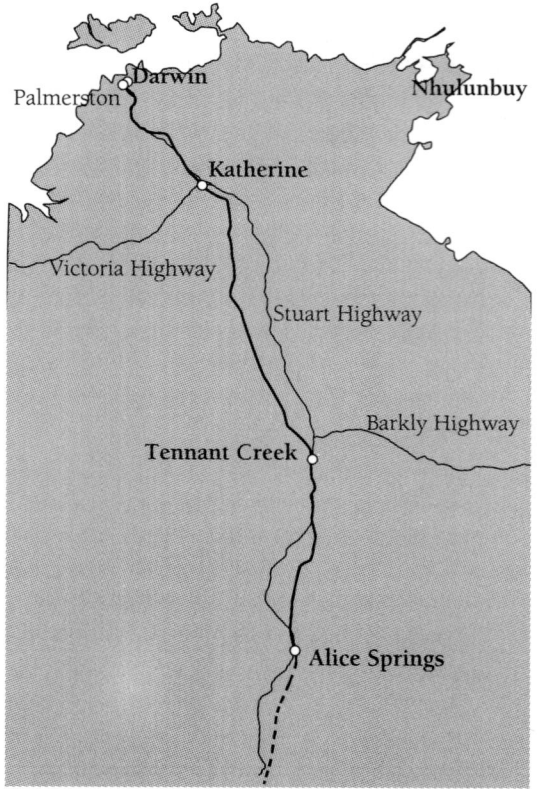

Source: http://www.clickforaustralia.com/mapoofnorthernterritory.htm
(accessed on July 8, 2009).

Furthermore, in developing countries, serving rural communities for their development is also a major function of universities, especially agriculture-oriented universities. In this consideration, many universities have been involved with rural development programs. In September 1998, there was an Asian and Pacific regional meeting organized by UNESCO and INRULED on the Role of Universities for Rural development. The participants shared information and also devolved arguments on the role of universities for rural development. For instance, Dr Paitoon Sinlarat (Thailand) in his paper, "Role of the Thai Universities in Rural Development: Time for New Concepts and Methods" reviewed that Thailand and Thai universities have developed in the field of agriculture and in the areas of rural development studies. "There is a need for a shift in the thinking of universities for new or alternative models of rural development with diversified approaches, which make proper use of local resources and knowledge for sufficiency while making profit as well" (Sinlarat, 1998 pp. 51–8).

Mrs Priscilla Cabanatan (SEAMEO INNOTECH [Southeast Asian Ministers of Education Organization-Regional Centre on Educational Innovation and Technology], Philippines) in her paper, "Contribution of Universities to Rural Agricultural Development through Science and Technology," argued that Asian countries generally see the development of the higher education sector as integral to their economic and social development plans. In this context, universities are seen as furthering national goals. They contribute to achieving such goals through their threefold functions of teaching, research, and extension. Her paper also gives some examples, drawn from developed and developing countries, on how universities have performed their functions so as to contribute to rural development. Examples with strong technology components have been emphasized to show that technology has great potential for enabling universities to maximize their contributions (Cabanatan, 1998, pp. 63–9).

Dr Darwin Karyadi (Indonesian) in his paper, "Poverty Alleviation Strategies: Indonesian Case," pointed out that the 21st century is a challenging time for new roles and functions of universities for rural development. Indonesia has a long-standing experience of tripartite co-operation between students, social communities, and universities in rural development, notably through the so-called SAP (Student Action Programme) in rural areas. The SAP has preset objectives and roles for each component: student, society, and local government, and enables the university's leadership in fostering better and harmonized partnerships to deliver appropriate systems to fulfill the basic human needs of the poor households such as food, health education, sanitation, shelter, income, etc. Another model established by SEAMEO (Southeast Asian Ministers of Education Organization)–University of Indonesia in the rural community, under the framework of the Poverty Alleviation Program with other development sectors, are the methodological approaches targeting the poor household, particularly vulnerable subgroups. The intervention design for the improvement of nutritional/health status, impact evaluation may contribute to further regional and international institution (Karyadi, 1998, pp. 91–96).

Dr Pracob Cooparat (Thailand) presented the paper, "Information Technology as Strategies for Rural Development," and emphasized that rural development is an issue of global concern; however, most colleges and universities in Southeast Asia are urban and biased not in favor for rural development. Now, it is more important than ever that most universities and colleges should have their strategies in contributing to rural development in their own possible ways. At present, traditional modes of teaching and learning in most universities require resources which are too expensive and have limitations to contribute to teaching and learning activities for the mass of rural people. For entering the 21st century, it is suggested that Information Technology should be used (Cooparat, 1998, pp. 77–85).

The Indian experiences shared by Dr K.V. Raman in his paper, "Role of Universities: Indian Scenario," highlighted that rural development is a comprehensive and multidimensional concept with an objective of improving the quality of life of the rural people. It can be achieved by the development of (a) agriculture and allied activities, (b) socioeconomic infrastructure, and (c) human resources through skill development and improvement. The university can efficiently perform the above rural development functions only through: (a) education by appropriate curricular involvement; (b) systematic studies and research on rural problems and constraints; (c) developing appropriate technologies in areas like health, sanitation, agriculture and animal husbandry, family welfare, education, employment rural industries, energy and resource use and management, and ensure that they are location-tested, adapted, and disseminated; (d) to act as "resource centers" for information by networking with other institutions. A novel experiment of a one-semester study for students in a farm and rural environment, known as "Rural Agricultural Work Experience" has been a successful experience in providing two-way interaction between farmers and students, exposing the former to modern science and technology while the latter learn the rural socio-economy and constraints (Raman, 1998, pp. 107–11).

Other experiences in India discussed by Prof. Ram Takwale in his paper, "Experience of and Expectations from Indian Universities," emphasized that the developments in IT should be used extensively to create distributed classrooms through interactive television, to transmit instructional materials, and to ensure intimate interactivity amongst students and teachers who could be living and working at a distance. He also proposed that a campus-based structure of the university should be changed into a decentralized university having its centers at or near villages. They should be well connected through electronic communication. Students and teachers should live/study/teach/train/work and learn at these centers along with villagers. The university campus should become a resource center for specialized facilities. Working and learning collaboratively and achieving development should be the methodology of the university. In this endeavor, the university should establish partnerships with agencies and organizations—governmental and nongovernmental—in carrying out its educational and developmental activities (Takwale, 1998, pp. 131–37).

Dr Mohammad H. Rahman (Bangladesh) in his paper, "The Case of Bangladesh," argued that clearly, universities can provide support at three levels: (a) policy support at the national level (e.g., advisory support, policy advocacy, and policy experimentation); (b) technical support at the subnational level (e.g., consultative support to research organizations on species selection, input and agricultural extension service selection, etc.); and (c) implementation support at the local level (e.g., Pilot scheme, innovative strategies, and linking students and researchers with agricultural/rural development projects and replicating the successful programs for wider implications). But, a somewhat generally agreed upon conclusion is that neither poverty alleviation nor sustainable development has been achieved to a recognizable degree. While other institutional problems are partly responsible for this, the lack of linkage with universities is also seen by many, as an important reason for Bangladesh's rural under development. Although there are good numbers of universities that can be linked with rural development efforts at government and non-government organization (NGO) fronts, the faculty resources have been largely unutilized for agricultural as well as rural development (Rahman, 1998, pp. 119–26).

Dr Yar Muhammad Khan (Pakistan) described "The role of the university in the rural development, in the context of 21st century," and expressed that to fix the role of university in the context of rural development in the 21st century, the university has to change its present mandate from mere academics and some sort of research and extension, and developing a more viable strategy for a holistic developmental process in rural areas and act as advisor, implementers, or partner in the implementation of rural development plans. The university on the one hand shall train and produce Graduate Catalysts (GCs) to live with the local community, to listen and learn from them, and help them in solving their problems and developing and executing plans. GCs should guide them toward self-reliance through the process of social mobilization, capacity-building, capital formation transfer of technology, skills enhancement, productive linkages, viable inputs, and delivery system. The university in the future has to have more elaborated integration and co-ordination with national building departments, the political lords, NGOs, private and public entities to have an integrated and collaborating working environment. To achieve the above-stated objectives, the university has to frame new instructional methodologies, programs to sensitize their students and teachers in social dynamics and operational paradigms. The university has to use the advanced information technology to update its working with communities and international resource centers. A new innovative rural development and educational policy would be a prerequisite (Khan, 1998, p. 145).

Dr Wan Hashim Wan Teh (Malaysia) described the Malaysian experience in his paper, "Role of Universities for Rural Development the Malaysia Experience." The story is through its village adoption scheme (VAS); approximately a hundred villages have benefited from new knowledge, ideas, technology practices, and innovation from the university's faculty members. The aim of the VAS is to empower self-help communities guided by such related principles as participation, diligence, group discussion, working together trustworthiness, and community responsibility. Development in this context is focused on the community members themselves who will plan and implement their own ideas while the university provides technical expertise and advisory services.

It is also suggested that effective rural development could be achieved through a smart partnership and close co-operation of three parties: the university, government agencies, and people where the relationship is mutually beneficial to one another (Teh, 1998, pp. 151–54).

As one of the developing countries, rural development was set up as a fundamental goal by Chinese universities. Apart from AUH's contribution for rural development, other universities in China also have engaged in the similar activities that can be summarized as following.

Mr An Ning (China Northwest Agricultural University) in his paper, "The Challenge and Expectations of China's Agricultural Education and Development," shared a view that agriculture is strongly affected by local conditions; therefore, agricultural education should be carried out in the light of local specific conditions and their distinctive characteristics. For the local community and the regional economy, the only way to better run the agricultural university is serving the agricultural, rural economy and farmers. As the intellectual economy approaches, agricultural universities should also accelerate its reform in the fields of teaching, research, and extension. In particular, agricultural science and technology, demonstration, and extension services need to be strengthened, so as to meet the needs of regional economic development, while the agricultural university will also benefit from the service (Ning, 1998, pp. 59–62).

Dr Luo Xiwen (South China Agricultural University) gave an example of how contributions by South China Agricultural University to the development of agriculture was through helping the government and the farmers prepare development strategies and participate in the development of the rural area; demonstrating the university's scientific research achievements for the farmers; running training courses and giving lectures on modern agricultural knowledge and techniques to farmers; co-operating with farmers' enterprises to transfer the university's scientific research achievements to farmers and enterprises and to develop a new way to contribute to education with research and production serving as a technical consultant for the development of farmers' enterprises to provide high-quality products (Luo, 1998, pp. 71–6).

Mr Guan Chunyun (China Hunan Agricultural University) presented the practices of Hunan Agricultural University to serve local economic development, which are readjusting the structure of specialities; deepening educational innovation and training qualified technicians in the construction of rural areas; carrying out scientific research closely around practical problems faced in rural production; and making efforts to turn research achievements into productivity (Guan, 1998, pp. 147–50).

Dr Li Xiaoyun (China Agricultural University) analyzed a gap between the current existing structure of the university setup and the realities of the rural development work. The criticism is that the current university setup is very much western-industrialization based/oriented and this needs to be first reviewed and changed in order to meet the needs of the rural population. The characteristics of the rural farmers of Asia (e.g., small-scale farmers are the predominant population) are very different from the rural population configuration of the western society (Li, 1998, pp. 139–43).

It can be concluded, therefore, universities can and do play important roles for rural development both in developed countries and developing countries, but different countries have different situations, requirements and considerations, and all those required the universities to have different strategies, principles, and priorities for their rural development programs.

Significance of This Study for Rural Development

"The past quarter century has been a period of unprecedented change and progress in the developing world" (Chambers, 1983, p. 1). Even though a great achievement in economic development and social progress have been obtained worldwide, more than one billion (1.115 billion) people in the developing countries continue to live in absolute poverty. Three quarters of them live in the countries of the Asia-Pacific region (Ordonez et al., 1998, p. 1).

With the biggest rural population in the world, China has made tremendous achievements in education, social and economic development in rural areas. The universities, especially agricultural universities, have played significant roles for rural development through their teaching, training, research, and extension work. Australia as one of the developed countries and probably among the smaller group of rural populations in the world has also paid attention to rural, remote, and indigenous communities.

With the view of sharing and exchanging experiences in the development of rural areas, the university and rural development were selected as the areas of study. This study will document how universities use their knowledge base to help rural transformation within an ecologically sustainable tradition with the view to developing a model of general principles that can be applied within a comparative perspective.

Though rural development is a goal of all kinds of societies, especially for developing countries, both the concepts of development, and the prevailing development models and policies have come under serious questioning during recent times. Attention has been drawn to structural adjustment policies that have severely affected social sectors such as education, health, agriculture, and so on (Ordonez et al., 1998, p. 4).

Rural development, "in any case, is not just a matter of altering economic growth rates and paths. It has to be viewed in much broader terms and encompass the fulfilment of each person's human potential in its material, spiritual, individual and social dimensions" (Ordonez et al., 1998, p. 4).

Before talking about and implementing rural development, the priority should be to give consideration to human resources development. The Human Development Report adopted by UNDP explained that "Human development is a process of enlarging people's choice. Three essential areas are for people to lead a long and healthy life, to acquire knowledge and have access to resources needed for a decent standard of living" (UNDP, 1991).

In practical terms, education is considered by governments, societies, and communities as a social instrument for developing human resources and increasing national productivity. And the university has played significant roles in this regard. This study will examine what different roles universities play in rural development; outline strategies of implementation done by these universities in implementing development in rural areas and thus compare these strategies made; determine strategies that are successful and unsuccessful; analyze what common features that make some strategies successful and others to fail; formulate a common strategy of action that can possibly generate more success in project implementations.

The two selected universities, one in Australia and the other in China, will be considered and compared. Universities considered in this study are included under the program of UNESCO-INRULED University-Link Program, which are: CDU, Australia and AUH, China; the two universities have played important roles of development in rural, remote, and indigenous communities in their own countries. By conducting a comparative study between these universities, methods and strategies can be outlined and analyzed as to their success/failures. From these outcomes, experiences and success can be shared, mistakes can be avoided, as well as common methods and strategy of action can be formulated so as to ensure a high probability of success in rural development which will replicate them in other areas with similar conditions. In doing this, there will be a wider exchange of expertise, knowledge, information, data, sharing, and learning. It is hoped that when the stories of two universities serving rural communities are developed, a comparative study method can be used and the "trial and error" approach to project development can become more efficient and effective.

Questions That Need Answering

How to promote development in rural areas? What would be the strategies needed to translate knowledge and technologies from universities into a project, such as an income-generating project (IGP) in rural areas? How can these activities be sustained so that when the project implementers leave the project site, residents will still continue to pursue the same activities for their own betterment without the necessary policing of project by the implementers? These questions have always bothered me since my involvement in some projects implemented in rural areas. In every phrase of project implementation, especially in income-generating programs, there have always been different sets of problems to handle. However, in many instances I have recognized that project implementers' visions cannot be sustained after several months when the project was terminated. Sometimes the termination comes earlier as the implementers leave the sites. Is it because, the people simply seemed to have no knowledge or skills? Or they do not have the required motivation or education thus to further increase their aspiration in life? It seemed to me that providing knowledge and technology alone without letting people know or understand what they will benefit from, does not prove worthwhile or cannot provide enough ingredients to increase people's motivation and aspirations to better their life. These questions keep on pressing me and give me the motivation to look deeply into methods and strategies for universities delivering rural development programs.

I understand that in any development project, a holistic approach must be taken. But the main question is what a holistic approach is. For me it seems that it is more easily said than done. In this regard, I advocate the importance of the process of documentation in all aspects of project implementation starting from the planning stage. In so doing, steps made can be reviewed and if something goes wrong the plan can be reconsidered and can be easily found out where the problem lies and immediate measure can be taken to correct it.

There are many factors that contribute to the success of rural development, for example, ownership, organization, the source of capital, the education level, government policy support, and the peoples' aspirations and motivations themselves. Which among these factors are more important for rural development? Are they interrelated to or independent of each other? With these questions in mind, the following chapters develop an understanding of the mechanism of rural development and transformation and the success factor affecting its sustainability.

2

The General Roles of Universities, Adult Education, and Agricultural Extension Education

Introduction

In the recent past, attention has been paid to how education and training has acted as a means to fulfilling rural transformation, rural economic growth, and social cohesion. This is related to deep understanding of the roles of universities, and their services in the knowledge and skills transformation in rural areas, adult education, and agricultural extension education.

Roles of Universities

Universities as higher educational institutions have many advantages in terms of human (staff and students) and intelligent resources for rural development, and they were, have been, and will be playing a positive contribution to development in this area. Universities should identify the real needs of rural people and have a responsibility to provide teaching and training opportunities for their development. Ping (1998) addressed that a university should be a place for the threefold functions of teaching, research, and extension/consultancy. For an agricultural university, all these should be closely linked with the rural human resources capacity building and rural development. Higher agricultural education has played a key role to ensure that knowledge and skills are transferred from teachers to students and other community members in rural areas. It has also helped the rural development sectors accept the role of agriculture and sustainable natural resources management (Atchoarena and Gasperini, 2003, p. 312).

Furthermore, according to the definition given by the 27th UNESCO General Conference (1993), higher education includes "all types of studies, training or training for research at the post-secondary level, provided by universities or other educational establishments that are approved as institutions of higher education by the competent State authorities." Thus, the core missions of higher education are to educate, to train, to undertake research, and to provide services to the community. The declaration of UNESCO Asia and Pacific Regional Conference on National Strategies and Regional Co-operation for the 21st Century (Tokyo, Japan, July 8–18, 1997) have given the aims of higher education as:

> Higher education is essential for any country to reach the necessary level of economic and social development and social mobility in order to achieve increasing living standards and internal and international harmony and peace based on democracy, tolerance and mutual respect. At the end of the Century, we reaffirm that the aims of higher education can be summarized as followings: to educate responsible and committed citizens, to provide highly trained professionals to meet the needs of industry, government and the professions; to provide expertise to assist in economic and social development, and in scientific and technological research; to help conserve and disseminate national and regional cultures, drawing on the contributions from each generation; to help protect values by addressing moral and ethical issues; and to provide critical and detached perspectives to assist in the discussion of strategic options and to contribute to humanistic renewal. (The Declaration of UNESCO Asia and Pacific Regional Conference on National Strategies and Regional Co-operation for the 21st Century, July 8–18, 1997)

The above paragraph outlines the roles of universities as a total enterprise, but in this book the following chapters will focus as a priority,
on the university role for rural development. Since "… [in] many
countries, higher institutions are heavily concentrated in urban areas,
whereas the majority of the population lives in rural areas, thus requir-
ing new mechanisms to address rural disadvantage" (The Declaration of
UNESCO Asia and Pacific Regional Conference on National Strategies
and Regional Co-operation for the 21st Century, Tokyo, Japan,
July 8–18, 1997).

In this book, I am looking for a comparative view between AUH,
China, and CDU, Australia, concerning their participation in rural communities for their development. The intention is to search for a model
which can be used in a more general application.

China is a developing country with a long tradition of Eastern
philosophy; Australia is a developed country with a modern Western
concept, therefore this comparative study will make a contribution as
"… [M]any leaders of higher education in the region see the need for
better integration of western concepts and values of countries with
Eastern philosophy and culture" (The Declaration of UNESCO Asia
and Pacific Regional Conference on National Strategies and Regional
Co-operation for the 21st Century, July 8–18, 1997).

Mr Jacques Delors, Chairman of UNESCO International Commission
on Education for the 21st Century, has mentioned in his report of
"Learning: The Treasure Within":

> Universities would contribute to the process of rural development by diver-
> sifying what they offer:
>
> 1. As scientific establishments and centres of learning, from where students
>   go on to theoretical or applied research or teaching;
> 2. As establishments offering occupational qualifications, combining high-
>   level knowledge and skills, with courses and content continually tai-
>   lored to the needs of the economy;
> 3. As some of the main meeting-places for learning throughout life, open-
>   ing their doors to adults who wish either to resume their studies or to
>   adapt and develop their knowledge or to satisfy their taste for learning
>   in all areas of cultural life; and
> 4. As leading partners in international co-operation, facilitating exchanges
>   of teachers and students and ensuring that the best teaching is made
>   widely available through international professorship (Delors, 1996).

Therefore, the universities, especially those in developing countries, should learn from their own past and analyze their countries' real situations and difficulties, and develop programs aimed at finding solutions for these problems. Universities are in a good position to provide vocational and technical training, adult training, and agricultural extension training for the rural community leaders, local technicians, and farmers. It is particularly important for the universities to find new development models which can be used in their own regions on a case-by-case basis, to enhance lifelong learning and rural development as these two concepts are closely connected.

> The concept of lifelong learning is of utmost importance. In rapidly changing economies, the labour market will constantly require new and different skills and so mechanisms must be enhanced to allow professionals to upgrade their skills at regular intervals and to develop new competences. People's needs of lifelong learning have expanded in all countries of the region. Higher education institutions thus offer learning opportunities in response to diverse demands and work co-operatively with other agencies and employers to ensure that appropriate courses are widely available. Ready access and flexibility in timing are of utmost importance. (The Declaration of UNESCO Asia and Pacific Regional Conference on National Strategies and Regional Co-operation for the 21st Century, Tokyo, Japan, July 8–18, 1997)

In his report "Learning: The Treasure Within," Mr Jacques Delors has identified four key functions of universities.

1. Preparing students for research and teaching.
2. Providing highly specialized training courses adapted to the needs of economic and social life.
3. Being open to all, so as to cater for the many aspects of lifelong education in the widest sense.
4. International co-operation.

A great deal of successful work in rural areas has been done by the universities through their training, research, and extension. University extension has not only served the community by its contribution to intellectual advancement and social progress, but it has also been of benefit to the universities themselves—by extending their influence and indeed their knowledge of the society that sustains them. More, it has promoted the multiplication of the universities themselves since many of them owe their foundation to the extension movement (Burrows, 1976).

Some of the critical considerations on the relationship between university and rural development could be summarized as: empowerment of the rural people to be self-sustainable; university's rural development programs could be accountable to the community; establishment of a broad network and partnership in rural communities; program integration; community participation and two-way communication; information technology and digital media; reform of existing structure and models, etc. (Final Report of UNESCO Asia and the Pacific Regional Meeting on the Role of Universities for Rural Development, INRULED, 1998, p. 18).

A sustained poverty alleviation program targeting the rural poor should focus on: (a) household food security; (b) nutritional security; (c) livelihood security; (d) economic security; (e) environmental security including health, sanitation and environmental management; (f) ecological security including bio-diversity consecution; (g) human resources development through skills; and (h) social and gender equity. The universities, as public service institutions, have a responsibility of subscribing to and implementing the "Declaration on Food and Nutritional Security" adopted at the World Food Summit (1996).

If universities are to play a constructive role in rural development, they have to adjust their programs to innovative and nontraditional topics, innovative teaching and learning models, as well as new partnerships with governments, academic persons, research institutes, and rural communities (Atchoarena and Gasperini, 2003, p. 321). Universities cannot succeed in participating in rural development programs unless they make their own contributions and establish successful models. Later Chapters will attempt to add both theoretically and practically, to this end.

Finally, to quote a sentence from "World Declaration on Higher Education for the 21st century: Vision and Action":

> Higher education should reinforce its community service, especially its activities aimed at eliminating poverty, violence, illiteracy, hunger and disease through an interdisciplinary and trans-disciplinary approach in the analysis of problems and issues.

Adult Education

What Is Adult Education?

There are many possible definitions and different aspects to adult educational practice in the world. And it is hardly unified. Different times, different countries, and different scholars have different opinions. Generally speaking, adult education is to educate the persons who missed some period of their education. Some individuals receive only a very incomplete education, and it is adult education's role to complement or substitute for elementary and professional education (Coles, 1977, p. xvii). Coles (1977) also states that adult education is for those whom it helps to deal with a new environment or requirement; it is also a further education to those who have already received training; conclusively, it is an educational activity aimed at individual development for everybody. Additionally, the meaning of the term, adult education, needs to be examined by looking at a number of statements that have been written about it. For example, some writers like Holster (1977) and Knowles (1990) suggest the education and training of adults and adult education are essentially the same. Others (Jarvis, 1983), explain that in the United Kingdom, the term adult education, specific meaning which imply liberal education, stereotyped as middle-class spare time education pursuits. While such a meaning to some extent existed in Australia, researchers concluded that adult education often reflects all educational activities of post schools (Shaw, 1999, p.15). The Australia Association of Adult and Community Education (AAACE) adopt the definition of adult education (AAACE, 1997), that is: "Adult education is any activity that deals with the education or training of adult."

The Chinese explanation of adult education can be summarized as follows: first, adult education should include both adult and education, which means adult education will provide systematic and continuing learning opportunity for an adult who has already held a role in the society in order to promote changes in personal knowledge, skills, attitude, and value; second, with a view of the functional meaning, adult education includes all organized education processes, regardless of the contents, levels and methods as well as formal, informal, and nonformal education. It is a replacement or extension of school education so that adults can acquire new knowledge and develop their potential; to promote professional qualifications and to change their thinking and behavior; third, adult education is a process, a social movement, a discipline, and a study area (Chen et al., 1999, pp. 17–18).

Adult education has been viewed as providing the part-time organized learning opportunities for the adult It is an educational process and it aims to promote change of personal knowledge, skills, attitude, habit, and value. Adult education covers literate education, continuing education, vocational education, and social and cultural education, as well as certificate education, and so on. It is one of the important components of lifelong education.

Adult education in Australia is also identified as technical and vocational education. It also includes some higher education courses (personal investigation). In China, adult education was known as education of workers and peasants or spare-time education, and only recently, it was officially called adult education, with an expansion of educational scope and diversification of educational reform (Adult Education in China, Chinese Ministry of Education, originally in Chinese, translated by Wang Li).

Adult education has been considered as a second chance for adults (Shaw, 1999, p. 15). Adult education both in Australia and in China is a significant area. The percentage of adults accessing education is considerable, as six out of ten Australians have had some form of educational experience as adults (Aulich, 1991) and in China, 76.93 million people became literate from 1987–98, and nearly 200 million since the founding of the People's Republic of China. The literacy rate for two-thirds of counties (cities) in China reached 95 percent or over. Illiteracy can hardly be found among the workers in cities and towns. Five hundred million participated in various forms of practical skill training from 1979 to 1998. Annual attendance of rural adult school training was over 70 million people in recent years (Adult Education in China, Chinese Ministry of Education, originally in Chinese, translated by Li Wang).

Adult Learning

Adult learning not only has formal education but it will also need nonformal and informal education as well, to cater to learners' different needs. Adult education is a way of providing access to knowledge for all.

It also helps people to understand the world and to understand others, and it offers an opportunity of learning and fulfilling one's potential. Adult learning is a period of learning in people's life and should be seen as a subset of lifelong learning (Sakya, 1993, p. 6).

Learning can be defined formally as the act, process, or experience of gaining knowledge or skills. Learning helps people move from novices to experts and allows them to gain new knowledge and abilities, to change attitudes, value, and to promote people's living standard. Delors (1996, p. 37) suggests:

> Education throughout life is based on four pillars: learning to know, learning to do, learning to live together and learning to be .... Formal education systems tend to emphasize the acquisition of knowledge to the detriment of other types of learning; but it is vital now to conceive education in a more encompassing fashion. Such a vision should inform and guide future educational reforms and policy, in relation both to contents and to methods.

What motivates adult learners? Typically, adults have different motivations for learning. Jeffrey A. Cantor suggests that some motivation for adult learners can be summarized as to make or maintain social relationships; to meet external expectations, for example, to upgrade some kinds of skills to keep a job; to learn to better serve others; professional advancement and the pure interest (1992, pp. 37–38).

Furthermore, there are different motivations for adults to keep learning. For instance, some might look for a better job; others might be wanted to have higher studies. Moreover, some might be interested in what they learnt, or what they wished to know. In rural areas, most adult learning contents are in agriculture, health, and social welfare, etc., which is closely linked with the adult's daily life and improvement in the quality of life.

The main aim for adult learning is to improve the quality of adult labors and make the better life (Coles, 1977). Through adult learning, the adults can change their mind, attitudes, values, and ability, and become confident in a society and in their own roles, as well as knowing what they want to do, and what to obtain. Mr J.K. Nyerere, the former president of Tanzania stresses:

> The education provided must therefore encourage the development in each citizen of three things; an enquiring mind; an ability to learn from what

others do, and reject or adapt it to his own needs; and a basic confidence in his own position as a free and equal member of the society, who values others and is valued by them for what he does and not for what he obtain (Nyerere, 1973, p. 247, as quoted in Coles, 1977, p. 11).

Coles (1977) further states that:

The three ingredients signify the liberation of man from ignorance, not to become a thought-less robot passively receiving and executing orders without dissent, but to be a creative, sensitive, aware, participating member of society, making the fullest contribution of which he or she is capable. (p. 11)

The field of adult learning was pioneered by Knowles (1950), who has identified the following characteristics of adult learners, or outcomes that adult learning should produce, that is: Adults should acquire a mature understanding of themselves; adults should develop an attitude of acceptance, love, and respect toward others; adults should develop a dynamic attitude toward life; adults should learn to react to the causes, not the symptoms, of behavior; adults should acquire the skills necessary to achieve the potentials of their personalities; adults should understand the essential values in the capital of human experience; adults should understand their society and should be skilful in directing social change (Knowles, 1950, pp. 9–10).

In adult learning, a range of critical issues have been developed. Ron Zemke and Susan Zemke (1988) suggested that "increasing and maintaining ones sense of self-esteem and pleasure are strong secondary motivators for engaging in learning experiences." He also mentioned that "new knowledge has to be integrated with previous knowledge; that means active learner participation". Brookfield (1986) shares the same opinion and emphasized that "prior learning experiences have the potential to enhance or interfere with new learning." "The effective adult learning entails an active search for meaning in which new tasks are somehow related to earlier activities" (Knox, 1977, as quoted in Brookfield, 1986). Unlike the formal learning, adult learning is a non-formal learning process. Some people stressed that "adult learning must be problem and experience centred" (Gibb, 1960, as quoted in Brookfield, 1986). Adults experience anxiety and ambivalence in their orientation to learning (Smith, 1982).

Brundage and MacKeracher (1980) have expressed how adult learning is facilitated as:

> Adult learning is facilitated when the teacher can give up some control over teaching processes and planning activities and can share these with learners; adult learning is facilitated when teaching activities do not demand finalized, correct answers and closure; express a tolerance for uncertainty, inconsistency, and diversity; and promote both question-asking and answering, problem-finding and problem-solving; adult learning is facilitated when the learner's representation and interpretation of his own experience are accepted as valid, acknowledged as an essential aspect influencing change, and respected as a potential resource for learning; adult skill learning is facilitated when individual learners can assess their own skills and strategies to discover inadequacies or limitations for themselves.

Some examples of literature suggest learners who are actively engaged in the learning process will be more likely to achieve success (Dewar, 1996; Hartman, 1995). When adult learners are actively involved with their learning process they begin to feel empowered and their personal achievement and self-directed levels can be raised. A key issue to getting adult learner actively involved in the learning process lies in understanding learning style preferences, which can positively or negatively influence learners' performance (Birkey and Rodman, 1995; Dewar, 1996; Hartman, 1995). It has also been shown that adjusting teaching materials to meet the needs of different learning styles benefits all learners (Agogino and Hsi, 1995; Kramer-Koehler, Tooney, and Beke, 1995).

It is apparent that adult learning is an educational activity, and it is an effective tool to affect change in attitude and behavior toward life; to empower adults' unharnessed potential so as to be capable to attaining the best in life.

Adult Education and Human Resources Development

Human development is much more than that the income is raised. It is a comprehensive development; it is creating an environment in which the people can have more potential, more productive and creative lives, therefore, human development is to enlarge the people's choices, to keep confidence, and to realize their values (UNDP Human Development Report, 2001, p. 9).

There was a special contribution for UNDP Human Development Report 2001 from Mr Kim Dae-jung, former President of the Republic of Korea. He said:

> We are living in an age of knowledge and information, fraught with both opportunities and dangers. There are opportunities for the underprivileged and poor to become rich and strong. But at the same time there is a danger that the gap between rich and poor nations could widen. The message is clear. We must continue to develop our human resources. The success or failure of individuals and nations, as well as the prosperity of mankind, depends on whether we can wisely develop our human resources. (p. 24)

In any society, if its members are educated and have high qualifications, capacity, and ability, even though they are still very poor, they are a "rich" society. There will be a great potential for their development. Mr Kim Dae-jung also shared the opinion as "If we succeed in developing the potential of our citizens by fostering a creative spirit of adventure, individual and nations will become rich, even if they are without much capital, labor or national resources" (UNDP Human Development Report, 2001, p. 24).

Education is an important vehicle to empower human beings and societies for their development and progress. It is clear that illiterate or functionally illiterate people cannot become the main forces for development of local economy and improvement of social progress in rural areas. It is also obvious that rural development cannot be realized without paying great attention to its human resources development. It is also undoubted that if the adult population, who are the main working force in any society, remain nonliterate, and unproductive, unresponsive to the changing environment of the world, the society will remain poor even though it is rich in natural resources. Therefore, adult education as a part of continuing education can increase the adult literacy rate, which is one vital human development index, so as to make a direct and significant contribution for human resources development, and also for economic development and social progress. Coles (1977) stressed that "real development must depend on the balanced growth of the person, both as an economic and social being." Furthermore, as quoted in Coles 1977, Harbison (1965, p. 71) emphasized that "the wealth of a country is dependent upon more that its natural resources and material capital; it is determined in significant degree by the knowledge, skills and motivation of its people." Singh (1998, p. 10) stated that "Education is a process of personal development through the harness of cognitive competencies."

Adult education is as an important part of lifelong learning and plays a significant role for human resources development and adult capacity building. It has been paid a great deal of attention by different countries and the international society. In 1997, UNESCO held the "Fifth International Conference on Adult Education" in Hamburg, Germany, and used "Adult Learning: A Key for the 21st Century" as its main theme. A hundred and fifty countries participated in the conference, and two important documents have been adopted, that is the "Hamburg Declaration on Adult Learning" and the "Agenda for the Future of Adult Learning," which focused on the roles of adult learning and lifelong education in 21st century. The important event of the follow-up activities was the "United Nations Week of Adult Learning" in September 2002 with the topic of "Build Learning Society: Knowledge, Information and Human Development," and took the "global dialog" as its initiative. All these efforts are intended to create a learning society.

Although the relationship between adult education and development is not simply considered as a cause and a result, and adult education cannot create jobs and automatically result in the individual or community development, adult education can empower the people and provide individuals more opportunities and abilities, as well as give them more potential.

Some theories and literature about adult development have emerged. Merriam and Caffarella (1999), as well as Clark and Caffarella (1999), indicate four models for adult development: biological model, which is concerned with how physical changes affect development; psychological model, which mentions that development is either sequential, a lifelong process or a series of transformation, or relational, a part of adult education; sociocultural and integrative models, which identify a new way of thinking about adult education for adult development. "Adult development is considered to be the transformation of individuals' existing knowledge to construct new knowledge as well as the reinforcement of existing knowledge. Changes are individually, socially, and culturally determined" (Billett, 1998, pp. 21–34). Dirkx (1998) summarizes four theoretical meanings on transformative learning in adults as consciousness raising, critical reflection, development, and individuation.

The three alternative views of adult education and development, such as the person centered view, the production centered view, and the problem solving view, were examined by Kuchinke (1999). Transformative learning in the adult is to make changes through transformation of the adult's perspective and meaning; to make senses of these changes frequently involves development (Dirkx, 1998, pp. 1–14). Daloz (1999) considers education as a transformational process and suggests that it is a way that adults can make meaning from their lives. Hobson (1998) carried out a study and indicates that "adult development, from a transformative viewpoint is more than adjustment to a particular society. It is a qualitative change in how the world is viewed and involves productive tension and struggle." Singh (1998, p. 3) has expressed:

> The changing world of work is a multifaceted issue of enormous concern and relevance to adult learning. Globalization and new technologies are having a powerful and growing impact on all dimensions of the individual and collective lives of women and men. There is increasing concern about the precariousness of employment and the rise of unemployment. In developing countries, the concern is not simple one of employment but also of ensuring secure livelihoods for all. The improvement needed in terms of production and distribution in industry, agriculture and services requires increased competencies, the development of new skills and the capacity to adapt productively to the continuously changing demands of employment throughout working life. The right to work, the opportunity for employment and the responsibility to contribute, at all ages of life, to the development and well-being of one's society are issues which adult learning must address.

Agricultural Extension Education

Agricultural extension education exists throughout the world in different forms as a means of disseminating useful knowledge about agricultural technologies to rural communities for the purpose of improving agricultural productivity and production, and bringing about change as well as improving the lives of farmers and their families. Agricultural extension education is a function of governments and specialized agricultural institutes, such as universities, research centers, or extension agents, because of the importance of agricultural production to a national economy, and the welfare of its people (Oakley and Garforth, 1985).

Agriculture extension education and development have long been a topic for discussion for many countries, especially developing countries. All extension activities in rural areas in developing countries take place within a process of development. The concept of rural development must be considered with particular reference to agriculture, since agriculture is a basis not only for national economy of developing countries, but also the livelihood of most rural families.

Development is an action or intervention process of social change. "It is a dynamic concept, which suggests a change in or a movement away from a previous situation" (Oakley and Garforth, 1985, p. 3). Rural development aims at changing rural people and rural society. It is not static but is continually evolving into new and different forms. Rural development introduces new ideas and new methods into rural society so as to improve people's living level. It is a process of transformation of a traditional society into an advanced one. During the changing or transformation, rural people can build a future for themselves and choose what to do by themselves. Rural development involves three elements: economic, social, and human.

> It should not concentrate upon one to the exclusion of the others. The economic base of any society is critical, for it must produce the resources required for livelihood. But we must also think of people and ensure their active participation in the process of development. (Oakley and Garforth, 1985, p. 3)

Agriculture is closely linked with rural development both for food and cash crops. It is an important economic activity for most countries in the world. In the developing countries, agriculture is usually the preeminent economic activity, and is crucial for overall economic development and social change. However, the vast majority of the people in developing countries live in rural areas, part or almost their entire livelihood comes from agriculture which has benefited little from advanced technology. Agricultural extension education has become a crucial tool for disseminating knowledge, technologies, information, and skills in rural areas for rural development.

Extension work is a means by which new knowledge, skills, and ideas are introduced to rural communities in order to bring about change and improvement in their livelihood. Without extension work, farmers would hardly be able to access the services for improving their agricultural production and other activities. Oakey and Garforth have pointed out in their book: *Guide to Extension Training, Food and Agricultural Organization of the United Nations*: "The critical importance of extension can be understood better if its three main elements are considered: Knowledge Communication, Farm Family" (1985, p. 7).

Extension is an informal educational process, taking the rural population as a target group to provide them with technology and information so as to help them solve their problems and empower their abilities for their own future development. Oakey and Garforth (1985) have concluded that "if the current ideas and practice of extension are considered, four main elements can be identified within the process of extension: Knowledge and skills, technical advice and information, farmers' organization, and motivation and self-confidence" (p. 8).

Extension is a process of training and learning, but it is different from normal university study, since most trainers are farmers who already have a lot of knowledge about their own areas and farming systems. From this point of view, extension education must be built on the existing knowledge and provide new ideas and information.

It is recognized that extension can play an important role for food security. Gasperini (2000) emphasized agricultural extension education for food security and pointed out "All for education and food for all." He also highlighted the idea of education is essential for empowering the poor and achieving food security. Sommer (2001) shared the same opinion and addressed "Education and Food for All," and rescripted the cases that contribute to the process of enabling rural people to improve their living and livelihoods. Bawden (1996), Crowder (1998), and Muny (1997) stressed the relationship between agricultural extension and sustainable development and expressed the integrating sustainable development theme into agricultural extension. Economic impact and contributions are critical issues for agricultural extension programs and rural development. Evenson (1997) reviewed and analyzed those issues in 75 studies (countries). Other literature has been found regarding agricultural extension education and rural people's welfare and rural sustainable development, such as Adhikarya (1995), Crowder (1996a, 1996b, 1996c), Deshler (1997), Farrigton (1997), and Zinnah (1998).

The world is changing. There are more challenges for agricultural and extension education, more demands of rural population for their development has emerged. Maguire (2000) emphasized that rural development is a complex process and many dimensions are considered with and referred to, such as sustainable production agriculture, natural resources management, institutions, infrastructure, health, education, markets, finance, policy, local government, and so on. In order to reach successful rural development, agricultural education coming from universities to nonformal adult education has to make changes to meet the expectations of the people and society. He also suggested that:

> Increasing competition from other educational institutions and non-traditional sources makes a strong and urgent case for agricultural education systems to make changes in order to influence a wide range of stakeholders including those in academia, in farming and non-farming rural areas, policy makers, and the private sector. (Maguire, 2000, p. 1)

It is time to have change. Innovative approaches and models should be identified that enable knowledge and technologies to be transformed for rural communities in order for them to make contributions for rural development.

Conclusion

The roles of universities, adult education, and agricultural extension education have been discussed with the view to develop an acceptable definition of each. All of these areas are relevant to the particular issues under discussion. Education and training, by equipping people with appropriate knowledge and skills and fostering values of human dignity, can expand rural people's choices and capabilities these choices.

3

Educational Development in Rural Hebei, China, 1949–2011

Introduction

Before going into the roles of universities in rural development, some important issues must be identified to understand the educational and social situations in rural Hebei. This chapter will discuss the educational, social, and demographical issues of rural Hebei, and where the AUH is located, as is necessary within comparative research methodology.

The review of experiences in China's social development and economic growth shows that education has made great contributions to the development in the rural sector. However, the relationship between progress in education and aspects of national development is not a cause-and-effect relationship, but an interactive one. It is in this consideration that the first inquiry will be made regarding the natural and socioeconomic contexts of rural education in Hebei Province. This inquiry will help in understanding the impact of social and economic factors on the role of education for rural population in rural development of the province.

It was decided that Hebei Province, China, would occupy a chapter; the NT, Australia, would occupy a chapter too; and then a concluding comparative summary would be made rather than attempting to compare and contrast each within one chapter. It is hoped that this approach will facilitate the reading and allow for a smoother summary analysis. In this way, it is hoped much repetition will be avoided.

The Social and Economic Context

Geographic Context and the Administrative Structure

Geographical Context

Hebei Province is in the northern part of the North China plain, to the north of the lower Yellow River and to the west of Bohai Sea with a 500-km-long coastline. It is between longitude 113°27' E.–119°50' E. and latitude 36° 3' N.–42° 4' N., bordering on Liaoning, Inner Mongolia, Shanxi, Henan, and Shandong Provinces. The Chinese name for "Hebei" means "to the north of the river," because the province, with Beijing, the capital of China in the centre, lies to the north of the lower Yellow River.

With the sea in front of it and mountains at its back, Hebei Province slopes down from northwest to southeast, and it is roughly divided into four parts, such as North Plateau, North Hebei, West Hebei mountainous regions, and Hebei Plain. The hilly area and plateau take up 57 percent of the total area of the province.

The plateau is 1,200–1,500 meters above sea level. The hilly land is an excellent natural pasture of Hebei. The Yanshan Mountain ranges in the north of it and the rolling Taihang Mountains ranges in the west, vary widely in height and have many high cliffs. As part of the North China Plain, the Hebei Plain is generally less than 50 meters above sea level. Its crisscrossing rivers and vast fertile land serve as an important grain and cotton growing area in the province.

The flat and low central part and the Binhai Plain contain lakes and rivers, for example, Baiyangdian Lake and Wenan depression. With most of their sources from the mountainous areas in the west and the north, the rivers all gather in the east before flowing into the Bohai Sea and form a water drainage system which mainly consists of Luanhe, Chaobaihe, Jiyunhe, and Haihe rivers. The Haihe River system comprises five branch tributaries including Yungding, Daqing, Ziya, Nanyun, and Beiyun rivers, with a total length of 1,100 km.

With a temperate continental monsoon climate, it is cold with a little snow in winter, hot and rainy in summer, windy and dusty in spring, and fine weather in autumn. The average annual temperature ranges from 1°C to 13°C. With the average annual temperature of 1°C, the plateau area, in the north part of Hebei, is the coldest part; the lowest temperature will be 40°C below zero. Handan in the south is the warmest, with an average annual temperature of about 13.6°C. The frost-free period lasts between 120 days and 200 days. Annual precipitation is 500 mm, and the rain falls mainly in July and August.

Hebei ranks 14th in land resources in China with a total area of 187,700 km² (http://www.linktrip.com/hebei/, accessed on April 5, 2010), but with its large population, the resources per capita are poor. There are over a thousand utilizable flora and fauna species that make up 25 percent of China's total. Hebei's marine resources are rich and also there are many locations that attract tourists (*Hebei Statistical Yearbook*, 2002).

Administrative Structure

China's governance is based on a four-level structure dividing the nation into provincial-, prefecture-, county-, and township-level administrative units.

The country (China) has 23 provinces, 5 national minority autonomous regions, 4 municipalities, and 2 special administrative regions directly administered by the Central Government. The Constitution specifically empowers the State to establish special administrative regions when necessary. A special administrative region is a local administrative area directly under the central government and free to maintain the capitalist system. At present, Hong Kong and Macao are the two special administrative regions of China (*China Statistical Yearbook*, 2009) (Figure 3.1).

A province or autonomous region is divided into cities, prefectures, and/or autonomous prefectures. A municipality or a city is divided into districts, counties, autonomous counties, and/or county-level cities (Figure 3.2). The district is the administrative unit covering the urban areas of municipalities and large cities. A prefecture or an autonomous

Figure 3.1
Administrative Structure of China

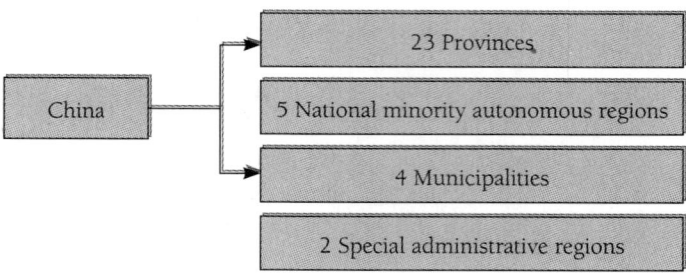

Source: China Statistical Yearbook, 2009.

Figure 3.2
Administrative Units of Hebei Province

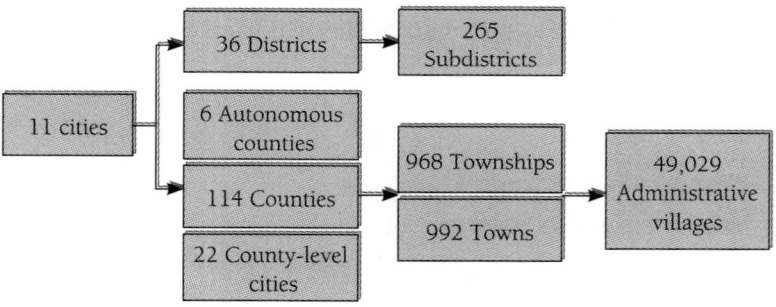

Source: Hebei Social and Economic Yearbook, 2010.

prefecture is subdivided into counties, autonomous counties, and/or county-level cities. A county, an autonomous county, or a county-level city is subdivided into townships, national minority townships, and/or towns.

In rural areas, the government has divided the big natural villages into the plains or grouped small ones in the mountains to form administrative villages, which are the basic unit of administration in China. An administrative village is managed by a village committee consisting of an elected chairman (usually called "village head"), a village accountant, a representative of the Women's Federation, a representative of public security, and the village technicians for agricultural and livestock production. The village committees are responsible for primary education and literacy training as well as other social services including family planning, food distribution, and public security at the grassroots level in rural areas. The size of a natural village (One administrative village may include a few natural villages.) is between 50 and 3,000 persons (China Social and Economic Statistical Abstract for Counties or Cities, 2009).

Economic Development

Present Economic Position of Hebei, China

Hebei Province belongs to China's "around Bohai Sea economic region." But amongst the fast developing eastern coastal provinces, the economic growth of Hebei is comparatively slow. In fact, Hebei is in the upper-middle level of economic development in China. (In this chapter, I have also used the data from the Jiansu Province, which is in the most developed region in China, and data from Gansu Province, which is one of the inland, remote, and less developed provinces. The data from above provinces could reflect the different development levels in China. The following data is the same.)

In 2009, the gross domestic product (GDP) of Hebei was valued at 1,723.548 billion Yuan, with per capita GDP being 24,581 Yuan. The per capita GDP, the per capita income of rural residents, the average salary of employees, and per capita consumption level of rural residents were all lower than the national average (see Figure 3.3).

The industrial sector of the province developed greatly after the 1950s. Hebei is currently a major industrial base for coal, steel, and textiles. Total provincial government revenue was 106.712 billion Yuan.

Figure 3.3
Economic Development Level of Hebei

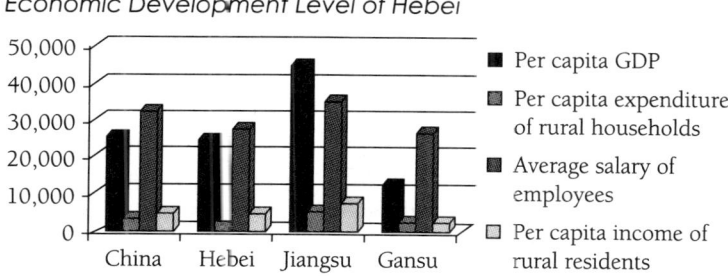

Source: Chinese Statistical Handbook, 2001.

Figure 3.4
Output of Grain, Cotton, Animal Husbandry, and Fishery in 2009

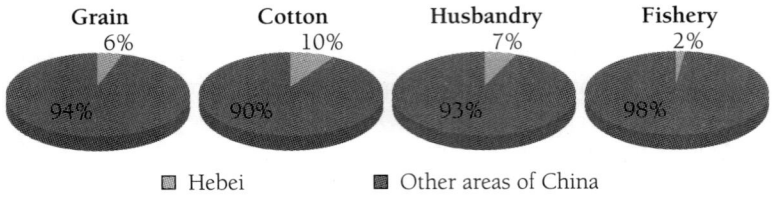

Grain	Cotton	Husbandry	Fishery
6%	10%	7%	2%
94%	90%	93%	98%

☐ Hebei   ▣ Other areas of China

Source: Data from Hebei Provincial Statistical Bureau.

Hebei has a long history of agriculture. It is one of the important areas for production of grain and cotton in China. Animal husbandry and fisheries are also important components of the economy. In 2009, the grain output of Hebei was 29.1017 million tons, with per capita yield of grain being 415.05 kg; the cotton output was 0.6064 million tons, with per capita yield of cotton being 8.62 kg. The animal husbandry product was 135.010 billion Yuan; the fisheries product 10.838 billion Yuan. The outputs of grain and cotton, and the values of animal husbandry and fisheries of Hebei accounted for an important proportion amongst the 34 provincial administrative units of China (see Figure 3.4).

Economic Development Before the Opening-Up Policy

The People's Republic of China was formally established on October 1, 1949, with its national capital at Beijing. Since the foundation of the People's Republic, among China's most urgent needs in the early 1950s were food for its population of 583 million, domestic capital for investment, and purchase of Soviet-supplied technology, capital equipment, and military hardware. To satisfy these needs, the government accelerated the redistribution of land in rural areas during 1951–52, which had actually begun under the Agrarian Reform Law of June 28, 1950, and began to collectivize agriculture. Preliminary collectivization was 90 percent completed by the end of 1956. In addition, the government nationalized banking, industry, and trade. Private enterprise in Mainland China was virtually abolished.

The period of "transition to socialism" began in 1953 and corresponded to China's First Five-Year Plan (1953–57). The period was

Figure 3.5
Growth of GDP by Type of Industry

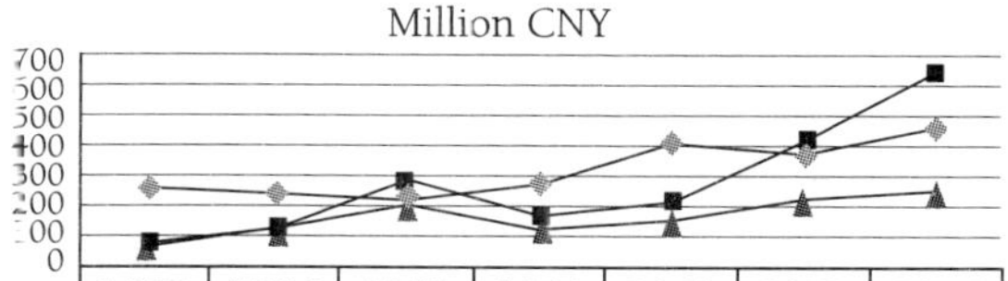

Million CNY

	1952	1956	1960	1964	1968	1972	1976
-♦- Primary industry (million CNY)	252.3	250.1	234.3	271.5	403.5	383.5	456.6
-■- Secondary industry (million CNY)	76.1	121.3	208.8	168.9	230.1	419.1	630
-▲- Tertiary industry (million CNY)	76.5	126.9	191.3	126.4	160.7	227.4	250.9

Source: Data from Hebei Provincial Statistical Bureau.

characterized by efforts to achieve industrialization and collectivization of agriculture.

In 1958, the Chinese Communist Party (CCP) launched the Great Leap Forward campaign under the new "General Line for Socialist Construction" principle. The Great Leap Forward was aimed at accomplishing the economic and technical development of the country at a vastly faster pace and with greater results. However, it was an economic failure followed by a five-year period of readjustment and recovery (1961–65) before the Great Cultural Revolution, which lasted 10 years and made negative influence on China's economic development again.

In this political context, economic growth in Hebei is comparatively slow (see Figure 3.5).

Economic Development After the Opening-Up Policy

Generally speaking, Hebei has experienced a rapid economic growth since the founding of the People's Republic of China. After the Cultural Revolution, Hebei entered a new era of development and the GDP increased at an annual rate of 10.6 percent. In 2009, the GDP of Hebei Province registered 17235.48 billion Yuan, ranking sixth in China (see Figure 3.6).

The per capita GDP was 24,582 Yuan in 2000, ranking the 12th in China (see Figure 3.7).

Figure 3.6
Growth of GDP in Hebei by Type of Industry

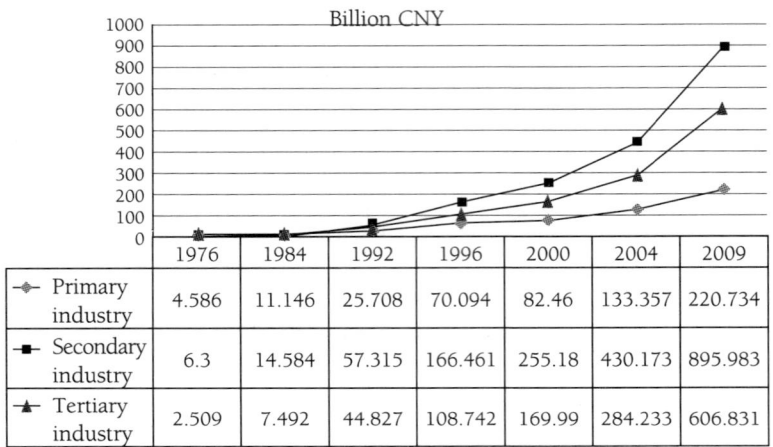

	1976	1984	1992	1996	2000	2004	2009
◆ Primary industry	4.586	11.146	25.708	70.094	82.46	133.357	220.734
■ Secondary industry	6.3	14.584	57.315	166.461	255.18	430.173	895.983
▲ Tertiary industry	2.509	7.492	44.827	108.742	169.99	284.233	606.831

Source: Data from Hebei Provincial Statistical Bureau.

Figure 3.7
Growth of Per Capita GDP in Hebei

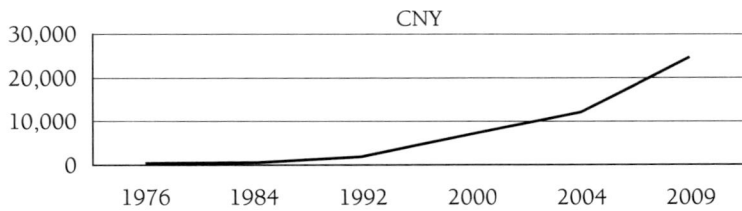

Source: Hebei Social and Economic Yearbook, 2010.

Economic development in Hebei has accelerated after "Cultural Revolution" and with the national opening door policy, the increasing rate kept of 10.6 percent annually. In 2009, it ranked 6th in China for total GDP and 12th in China for per capital GDP.

Demographic Context

China's Fifth Largest Populated Province

In 2009, the total population of Hebei Province was 70.34 million, China's fifth largest, and the population density was 375 persons/km^2. The density is higher in the south and central parts, lower in the north and west, higher in the plains, and unevenly distributed in the mountainous and hilly areas. The density is highest along the Beijing–Guangzhou railway line and Beijing–Shanhaiguan Highway. Generally speaking, the rural population is larger than the urban population.

By the end of 2009, Hebei had a employed population of 37.9349 million, 53.92 percent of the total provincial population. The wages of staff and workers totaled 133,792 million Yuan; total social insurance and welfare funds of employed and retired staff and workers were 12.21 billion Yuan. The per capita net income of rural residents was 5,149.67 Yuan. The average wage of staff and workers was 28,383 Yuan per year. The average household consumption was 7,193 Yuan, 3,606 for rural residents, and 12,195 for urban residents. The ratio of hospital beds per 10,000 persons was 23.30, and the ratio of doctors per 10,000 persons was 11.44.

In China, people are officially categorized into agricultural and nonagricultural status. (It was called "household registration system." Since the 1950s, every citizen had been assigned as a permanent personal status as a resident of the locality where he or she was born.) In rural areas, all the people, except officials and teachers employed by the government, are labeled and registered officially with an agricultural status. The percentage of agricultural population of Hebei was higher than the average of China (see Table 3.1 and Figure 3.8).

Table 3.1
Proportion of Agricultural and Nonagricultural Population in Hebei (percent) in 2009

		Total	City	County	Town
National	Nonagricultural	46.59	41.34	13.04	19.86
	Agricultural	53.41	58.66	86.96	80.14
Hebei	Nonagricultural	43.00	38.43	8.87	14.24
	Agricultural	57.00	61.57	91.13	85.76
Jiangsu	Nonagricultural	55.60	37.39	16.70	24.65
	Agricultural	44.40	62.61	83.30	75.35
Gansu	Nonagricultural	32.65	46.21	8.49	29.6
	Agricultural	67.35	53.79	91.51	70.24

Source: Data from *China Statistical Yearbook*, 2010.

Figure 3.8
Urbanization Status: Proportion of Nonagricultural Population (by Type of Administrative Unit in 2009)

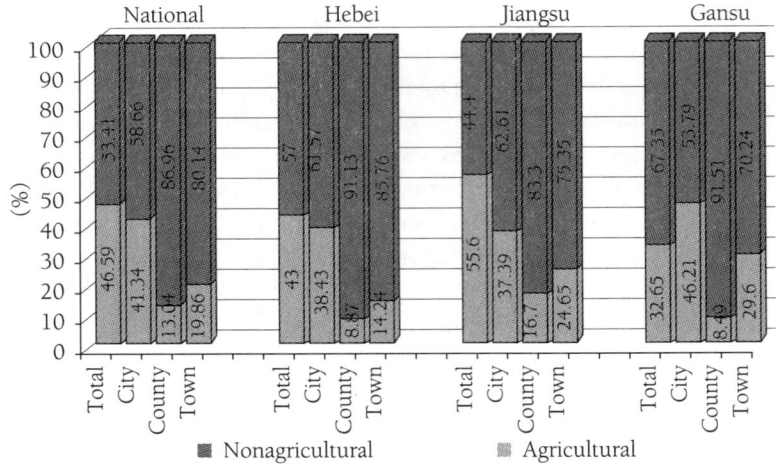

Source: Data from *China Statistical Yearbook*, 2010.

Figure 3.9
Urbanization Trend: Percentage of Agriculture Population in Hebei

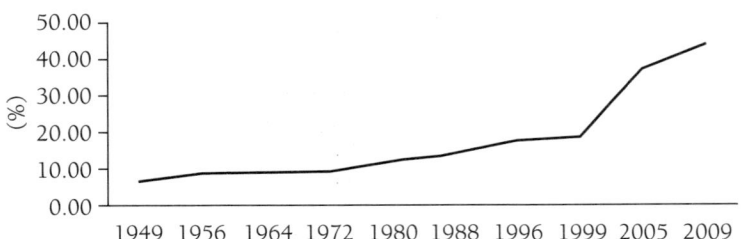

Source: Data from Hebei Provincial Statistical Bureau.

After the founding of the People's Republic of China, the nonagricultural population had increased by 12.22 percentage points in 60 years, from 6.76 percent in 1949 to 43.74 percent in 2009 (see Figure 3.9).

In fact, many people registered as agricultural population have migrated to the cities and lived in urban areas in the last two decades since China's reform and opening up to the outside world. According to population change projection, in 2009, emigration population in Hebei Province accounts for 53.09 percent of the total inter-migrants

Figure 3.10
Composition of Rural Employment in Hebei by Type of Industry

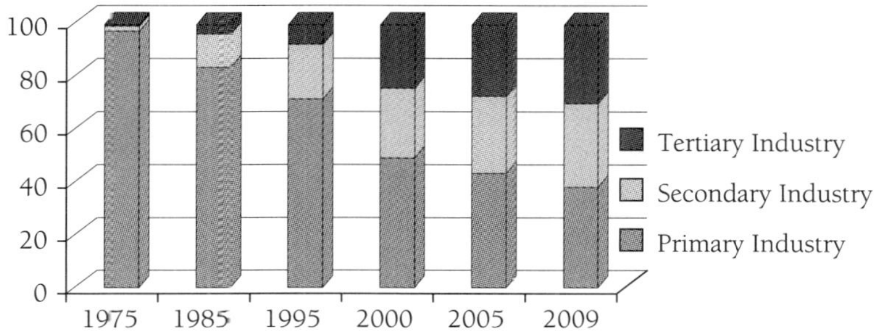

Source: Data from Hebei Provincial Statistical Bureau.

(*Hebei Economic Statistical Yearbook*, 2010). The main reason for emigration is to find a job or do business; the other reasons are job transfer or living with relatives. Finding a job or doing business, study or training, and marriage are the main causes of emigration. Marriage, job transfer, and following the family are the main reasons for immigration from other provinces.

The agricultural work force has become the main force of economic activities in both urban and rural Hebei. In 2009, 60.68 percent of the total work force with agricultural status was employed by nonagricultural sectors, for example, manufacturing and service industry (see Table 3.2).

Among the work force with agricultural status that stays in rural areas, more than 50 percent are involved in the secondary (in China, we categorized agriculture, manufacture, and service as primary, secondary, and tertiary industries) and tertiary industries (see Figure 3.10).

The Hebei population is characterized by an increasingly educated population and a decreasing rate of illiteracy.

Fertility Level and Changes

According to population projections in three scenarios, high, medium, and low the population of Hebei still continues to increase during the first decade of the new century with the natural increase rate recorded to be 5.7 per 1,000 from the year 2000–2010 (Figure 3.11).

Mortality decreased sharply since the establishment of People's Republic, much earlier than the decrease in the birth rate. The mortality rate dropped from 12.73 per 1,000 in 1949 to 6.43 per 1,000

Table 3.2
Employment of Rural Labor in Hebei by Type of Industry

		1975	1985	1995	2000	2005	2009
Primary industry	Employment	16,176,100	16,390,300	17,154,200	16,781,200	15,647,200	14,792,200
	%	97.74	83.83	72.24	49.56	43.84	39.00
Secondary industry	Employment	374	2,491,300	4,828,700	8,869,900	10,435,600	12,033,600
	%	2.26	12.74	20.34	26.20	29.24	31.41
Tertiary industry	Employment	0	671	1,762,300	8,206,000	9,606,900	11,099,100
	%	0	3.43	7.42	24.24	26.92	29.27
Total		16,550,100	19,552,600	23,745,200	33,857,100	35,689,700	37,924,900

Source: Data from Hebei Provincial Statistic Bureau.

Figure 3.11
Education Structure of Rural Work Force in Hebei (2009)

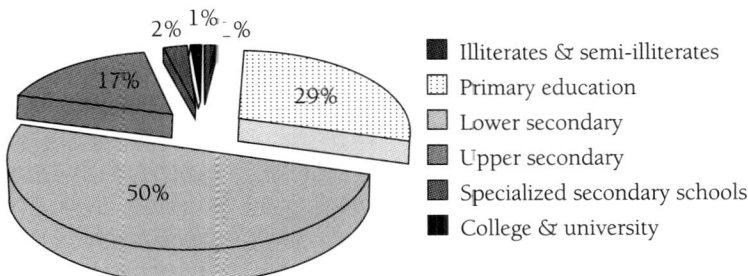

Illiterates & semi-illiterates
Primary education
Lower secondary
Upper secondary
Specialized secondary schools
College & university

Source: Data from *Hebei Agricultural Statistical Yearbook, 2010.*

Figure 3.12
Natural Changes of Population in Hebei

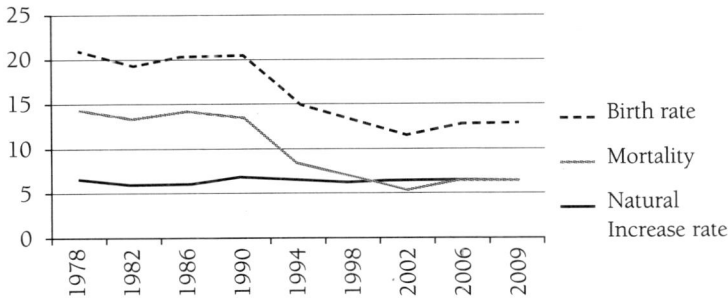

Source: Data from Hebei Provincial Statistical Bureau.

in 2009, remaining below that of national average. The infant mortality rates experienced a great decrease. Currently, the male mortality rate is higher than that of females.

The average life expectancies during 1929–33 were 40.03 years for males and 35.76 for females in Hebei; in 1990, they were 70.01 and 73.60 years, respectively, or an average of 71.70.

Hebei experienced a mortality transition from high mortality and low life expectancy before 1949 to low mortality and high life expectancy, close to the levels of developed countries (see Figure 3.12).

The population of Hebei has more than doubled since the 1950s. The process can be divided into five periods.

The period 1950–57 was one of rapid increase of the total population with an annual rate of 2.19 percent, or about 730,000 annually; in

Figure 3.13
Population Growth in Hebei

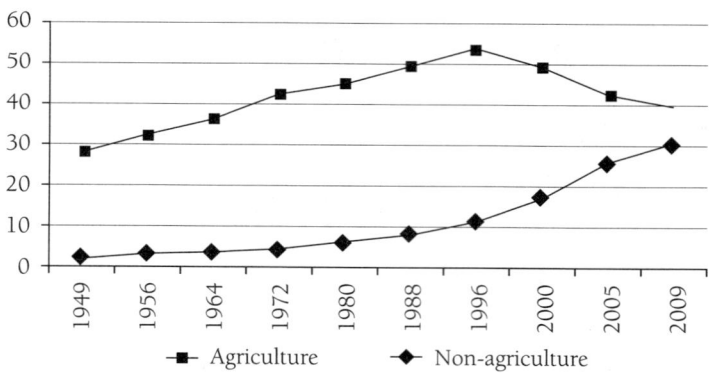

Source: Data from Hebei Provincial Statistical Bureau.

1958–61, there was a trough in population increases; in 1962–72, population growth bounced back and a peak in fertility was observed. The net population increase during the 11-year period was 9.3248 million, the average being 847,700 annually; in 1973–81, the speed of increase slowed down to the annual rate of 1.33 percent, or 587,400 annually.

In the fifth period, from 1982 to the present, population growth swung back with 597,741 additional people annually; the net increase during the last 29 years was 17.3345 million (see Figure 3.13).

Aging of the Population, Population Structure by Sex and Age

Since 1949, the sex ratio (F:M) of Hebei has increased gradually and smoothly to around 102 to 106 (M=100) on the whole. It was 104.03 in 1997 for the provincial total population. However, there is an uneven distribution of the sex ratios: Yanshan Mountain and Taihang Mountain have high sex ratios. This means that there are more females than males.

In 2010, out of a population of 71.854202 million, the population aged 0–14 was 12.093041 million, or 16.83 percent of the total; the population aged 15–64 was 74.93 percent of the total (53.841435 million); and the population aged 65 and above was 5.919726 million, or 8.24 percent of the total. In 1997, the total dependence ratio was 47.73 percent, with the young age dependence ratio being 37.63 percent and the old age ratio 9.80 percent.

According to the 2000 census, the population aged 60 and older reached 0.685954 million in Hebei, comprising 10.36 percent of the total population. The aged population increased rapidly, and is still increasing. The median age and life expectancy of the elderly are rising, and the dependence ratio for the aged population is also rising, which makes the dependency burden heavier for the working age population.

As for the distribution of the aged population, there were more people in the younger age groups: those in the 65–69 age group were 27.53 percent of the total, those in the 70–79 age group comprised 32.55 percent of the total, and people who were 80 years and older made up 8.52 percent of the total aged population. Mostly (72.52 percent) of the aged population were illiterates or semi-literates. Most of this aged population had been married and their employment rate was 29.78 percent.

Summary

This section has dealt with the social, economic, and demographic contexts of rural Hebei, based on statistical data. It is important for a comparative study to have a general background about this information, which could lead to a better understanding of how these contexts influence the education system in rural areas.

Development of Rural Education

Introduction

This section presents the experiences obtained and lessons learnt in the effort to provide education for rural population in Hebei Province since the founding of the People's Republic of China in 1949. The current structure of the educational system in China appears simple as compared to that of the 1950s. The general system of education comprises four stages: the primary, the junior secondary, the senior secondary, and higher education.

In terms of access to education, China's system represents a pyramid; because of the scarcity of resources allotted to higher education, student numbers decrease sharply at the higher levels. Although there were dramatic advances in primary education after 1949, achievements in secondary and higher education were not as great. By the end of 2009, there were 109 institutions of higher education in Hebei Province, with number of students enrolled being 332,278 and teachers 58,394; 3815 secondary schools with 1,382,740 students and 290,746 teachers; 14,447 primary schools had 875,800 students enrolled and 321,200 teachers (data from Hebei Provincial Statistical Bureau).

During the process of rural transformation, rural education refers to education for the rural population, including the children of the few government employees who have nonagricultural status. In 2009, population labeled as agricultural status accounted for 56.25 percent of the total population of Hebei Province. Provision of education service to rural population is still the main educational effort in Hebei Province.

Rural China usually provides only the first three stages, which have been reestablished at present as a 6-3-3 system (six year primary, three years junior secondary, and three years senior secondary) in general education in Hebei Province.

Children usually enter primary school at seven years of age for six days a week. The two-semester school year consists of 9.5 months, with a long vacation in July and August. Urban primary schools typically divide the school week into 24–27 classes of 45 minutes each, but in the rural areas the norm is half-day schooling, more flexible schedules, and itinerant teachers. The primary-school curriculum consists of Chinese, mathematics, physical education, music, drawing, and elementary instruction in nature, history, and geography, combined with practical exposures around the school campus. A general knowledge of politics and moral training through courses named communist ideology and morality, which stress love of the motherland, love of the party, love of the people, and love of the environment are another part of the curriculum. Chinese and mathematics accounts for about 60 percent of the total scheduled class time where as natural science and social science accounts for about 8 percent of it. *Putonghua* (mandarin) (http://memory.loc.gov/frd/cs/china/putonghua, accessed on January 6, 2010) is taught in regular schools and Romanized pinyin in lower grades and kindergarten. Most schools have after-hour activities at least one day per week—often organized by the Young Pioneers—to involve students in recreation and community service.

The regular secondary-school year usually has two semesters, totaling nine months. The academic curriculum consists of Chinese, mathematics, physics, chemistry, geology, foreign language, history, geography, politics, physiology, music, fine arts, and physical education. There are 30 or 31 periods a week in addition to self-study and extracurricular activity. Thirty-eight percent of the curriculum at a junior secondary school is in Chinese and mathematics, 16 percent in English. Fifty percent of the teaching at a senior secondary school is in natural sciences and mathematics, 30 percent in Chinese and English (China Statistical Data in Education, 1993–2000).

The system of rural secondary education has undergone several transformations since 1980. In each of the counties, a comprehensive center for vocational education and training has been established on the bases of schools and short-term training classes established by different government departments, businesses, trade unions, professional societies and clubs, and other organizations have been undertaken to reduce the administrative cost.

For the purpose of convenience, the section divides the development of education in rural Hebei into five major periods:

1. The Transition to Socialism, 1949–57;
2. From Great Leap Forward, 1958–60;
3. Readjustment and Recovery, 1961–65;
4. Cultural Revolution, 1966–76; and
5. Post-Cultural Revolution Reforms and Opening-up, 1976–present.

The Transition to Socialism, 1949–57

Before 1949, there were 32,484 primary schools and 173 secondary schools (50 general and 123 technical) in Hebei Province, and almost no adult education in rural Hebei except in the liberated areas led by the CCP and the Dingxian County rural education experiment led by Dr James Yen (Y.C. James Yen is a Chinese scholar, got the PhD degree from Yale University and served in China until 1949. He is also a founder of International Institute for Rural Reconstruction, Manila, Philippines). In November 1949, to answer the call of the education department of the People's Government of Hebei Province, 7,608 winter literacy classes were organized in rural Hebei to offer courses on basic literacy, and numeracy for 902,025 peasants. In 1951, literacy classes were established in 70 percent of the province's villages. In 1956, peasant schools were established in 80.8 percent of the province's villages on the bases of literacy classes (Data from Hebei Provincial Statistical Bureau).

At the primary level, the new government mobilized the villagers to establish a primary school in each village. Priority was placed on quantitative expansion of primary education to solve the problem of access to primary education in rural areas and realize the idea of equal opportunity in education.

Secondary education in China has a complicated history. The secondary school was stratified into six types as enumerated below.

Great importance had been attached to both general education and vocational education. The collectives are mobilized to share the educational burden with the government to solve the problem of deficiency of financial resources. Secondary teacher training schools were established to meet the needs of rapid expansion of primary schooling. In 1957, every county of the province had established at least one secondary school (see Table 3.3).

During this period, general secondary schools, technical secondary schools, and secondary teacher training schools were developed. Winter literacy classes were established in rural areas, some of which were developed into peasant schools. Moreover, accelerated-learning secondary schools for peasants were also established, and graduates could be promoted to higher education institutions to continue their education. Furthermore, regular full-time remedial schools for officials were set up to alleviate illiteracy among officials.

From Great Leap Forward, 1958–60

In 1958, the CCP launched the Great Leap Forward campaign under the *General Line for Socialist Construction*. The Great Leap Forward was aimed at accomplishing the economic and technical development of the country at a vastly faster pace and with greater results. Although the party leaders were to some extent satisfied with the accomplishments of the First Five-Year Plan, but Chairman Mao Zedong and his followers in particular believed that more could be achieved in the Second Five-Year Plan (1958–62) if the people could be ideologically aroused; and if domestic resources could be utilized more efficiently for the simultaneous development of industry and agriculture. The Great Leap Forward centered on a new socioeconomic and political system that created the people's communes in the countryside. The Chinese people started to embrace the slogans of "Long live the three banners of the Great Leap

Table 3.3

Development of Education in Hebei Before the Great Leap Forward Movement

Item	Year	Primary Education	General Secondary Education				Higher Education
			Total	General	Technical	Teacher Training	
No. of schools	1949	32,484		50			5
	1957	40,337	933	847	37	49	6
Enrolments	1949	2,265,621		20,501			1,623
	1957	4,122,907		372,567	22,982		8,513
No. of teachers	1949	55,483		1,150			287
	1957	102,610		14,106			1,217

Source: Data from Hebei Provincial Educational Commission.

Forward, the General Line, and the people's commune." The "People's Commune is the golden bridge that leads to the paradise in the world—the Communist society" and "The stronger the will of the people, the greater the soil will produce."

The individual commune was placed in control of all the means of production and was to operate as the sole accounting unit; it was subdivided into production brigades (generally coterminous with traditional villages) and production teams (i.e., work groups). Each commune was planned as a self-supporting community for agriculture, schooling, marketing, administration, and local security.

During the Great Leap Forward campaign, more and more rural children finished primary education due to the establishment of primary schools or learning posts (extension of complete primary schools in remote villages) in their villages. The development of secondary education for rural children became an urgent task. In 1958, the State proposed "to establish enough agricultural secondary schools, industrial secondary schools, and handicraft industrial secondary schools" to meet the needs of rural development. In 1964, Hebei Province started to practice a new educational policy of "dual educational system, dual labor system" proposed by the central government. The number of agricultural secondary schools, which were not preparatory for higher education, but to train an educated labor force for rural development increased rapidly in Hebei Province. In this type of school, courses on agricultural technology were added to the curriculum (see Table 3.4).

Table 3.4 shows that during the period of the Great Leap Forward, which was in 1958–60, more schools and universities in Hebei had been established in terms of secondary teacher training education, general secondary education, and higher education.

Table 3.4
Growth of Secondary and Higher Education in Hebei during the Great Leap Forward

	Secondary Teacher Training Education		General Secondary Education		Higher Education	
Year	1957	1960	1957	1960	1958	1960
No. of schools	49	124	847	1,375	6	94

Source: Data from Hebei Provincial Educational Commission.

Readjustment and Recovery, 1961–65

In 1961, in an effort to stabilize the economic front, the party, under the dominant influence of Liu Shaoqi and Deng Xiaoping, and others, a series of corrective measures were initiated. Among these measures was the reorganization of the commune system, with the result that production brigades and teams had more say in their own administrative and economic planning.

In this context, Hebei Province began to carry out the policy of "readjustment, consolidation, enrichment, and improvement." In 1963, because of deficiency of government financial resources, some schools administrative management got transferred from public to private mechanism. Consequently, 20 percent of the "gongban" schools (public schools funded by the State) switched to "minban" schools (rural schools funded by the collectives).

The policy of "dual educational system, dual labor system" was criticized and all the agricultural secondary schools were transferred into general secondary schools. By early 1965 the country was well on its way to recovery under the direction of the party apparatus, or, to be more specific, the Central Committee's Secretariat headed by Secretary General Deng Xiaoping.

In the early 1960s, educational planners followed a policy called "walking on two legs," which established both regular academic schools and separate technical schools for vocational training.

The government of Hebei carried out the policy of "walking on two legs" sincerely during the adjustment and recovery period after the great Leap Forward movement and readjusted its secondary and higher education (see Table 3.5).

As a result of readjustment, the number of secondary agricultural schools increased almost 10 times from 375 in 1962 to 3,625 in 1965 (see Table 3.6).

Peasant spare-time education also entered a new phase in 1964, in which peasant education began to shift from literacy through political study to learning agricultural technology according to local productive needs. For example, agricultural technology learning groups were organized in grain and cotton planting areas; electronic and mechanical groups were established in areas where agricultural machines and electricity were used; fruit tree planting technology groups were established in fruit planting areas; Livestock veterinary medicine groups were

Table 3.5
Readjustment of Secondary and Higher Education in Hebei Province

	Secondary Teacher Education			General Secondary Education		Higher Education	
Year	1960	1963	1965	1961	1965	1961	1962
No. of schools	124	22	25	1,375	897	94	45
Enrolments	69,298	7,543	12,605		422,328	60,000	40,000
No. of teachers	3,523	3,364	808		20948		

Source: Data from Hebei Provincial Educational Commission.

Table 3.6
Development of Secondary Agriculture Education in Hebei

Year	1962	1963	1964	1965
No. of schools	375	261	791	3,625
Enrolments	29,749	19,925	55,716	195,570

Source: Data from Hebei Provincial Educational Commission.

established in mountainous areas and the pasture areas in the plateau; and accounting and hygiene learning groups were established in all of the above named areas.

Cultural Revolution, 1966–76

On August 8, 1966, at the 11th plenum of the Eighth CCP Central Committee, the scope and strategy of the Great Proletarian Cultural Revolution was defined, and once again it was proclaimed that education had been controlled by bourgeois intellectuals, and that the creation of a new system more closely based on Chairman Mao Zedong's teachings was needed.

During the next three years, campuses were controlled in turn by propaganda teams of Red Guards, soldiers from the People's Liberation Army, and finally workers and peasants.

The primary schools were the least affected by the Cultural Revolution, and by August 1967 most had reopened for normal operation. However, primary education was, for the most part, shortened from six years to five years. The concept of Key Schools (in each city and/or county there are a few good schools in which students can have a better achievements) was abolished, with enrolments in primary and secondary schools based on proximity. In 1968, the Provincial Revolutionary Committee (replacement of Hebei provincial government during the Cultural Revolution) gave orders that all the gongban primary schools (public schools funded by the State) in rural Hebei should be transferred down to the productive brigades as minban schools (rural schools funded by the collectives). Since 1970, in order to reach the goal that primary schooling should be accessible within the production team, junior secondary schooling should be accessible within the production brigade, and senior secondary schooling should be accessible within the people's commune, primary and secondary schools expanded rapidly in rural areas. The rapid expansion of primary and secondary education during the Cultural Revolution, however, created serious problems; because resources were spread too thinly, educational quality declined. Further, this expansion was limited to regular secondary schools; technical schools were closed during the Cultural Revolution because they were viewed as an attempt to provide inferior education to children of worker and peasant families. The curriculum was reconstructed so as to conform to practical needs, resulting in the elimination of subjects such as history and geography. Even such core science subjects as physics and chemistry gave way to courses in industrial skills, biology to courses in farming skills (In 7, 2001).

In June 1966, the system of university entrance examinations was halted. However, few colleges and universities admitted new students until the early 1970s, and the selection of students was based on political virtue. Those from families of workers, peasants, or soldiers were deemed the most "virtuous," and were among the first admitted. This has generated the label of *worker–peasant–soldier student* for those students entering college during the early 1970s.

In all, the period of the Cultural Revolution was a very disruptive one for the Chinese society in general and its education in particular. The educational infrastructure in Hebei was decimated as a result of the revolutionary struggles, and students suffered because of a vastly watered-down or nonexistent curriculum. Perhaps the only gain, again at the expense of quality, was the delivery of primary education to an unprecedented percentage of school aged-children, largely because agricultural collectivization allowed for the creation of large numbers of "commune schools, overseen directly by the collectives rather than by higher level agencies (see Table 3.7).

Retaining literacy was as much a problem as acquiring it, particularly among the rural population. In 1966, peasants were involved in the political movement and no one attended literacy classes. Literacy rates declined between 1966 and 1976. Political disorder may have contributed to the decline, but the basic problem was that the many Chinese ideographs can be mastered only through rote learning or in verbatim and are often forgotten because of disuse. In 1968, literacy classes in rural Hebei enrolled 0.42 million learners. In 1972, the peasant schools in Hebei transferred into political evening schools.

Post-Cultural Revolution Reforms and Opening-Up, 1976–present

Reform of Educational System

The provision of basic education for all in so vast a country as China was a formidable accomplishment. Modernizing China, however, was tied to modernizing education. Decentralization of educational management from the central to the local level was the means chosen to improve the education system.

Rural secondary education has undergone several transformations since 1980. In 1982, the government of Hebei Province initiated the reform of the administrative system of education in rural Hebei and started to transfer the responsibility of primary and junior secondary schools in rural areas to the local government. The junior secondary schools were transferred to the people's communes and the primary schools to the production brigades. After the decentralization of the educational administrative system in the province, education boards were established at commune and production brigade levels to mobilize educational funds, improve educational infrastructure, and the physical conditions of rural education. In 1983, the communes were eliminated. In 1985, educational reform legislation officially placed rural secondary schools under local administration. The reform guaranteed quality primary education for all in rural Hebei.

Since the fall of the "Gang of Four," China entered the new era of post-Cultural Revolution reforms and opening up to the outside world. Among the notable official efforts to improve the education system was a

Table 3.7
Quantitative Expansion of Basic Education during the Cultural Revolution

Year	Primary Education 1965	Primary Education 1976	General Secondary Education 1965	General Secondary Education 1976	Secondary Teacher Education 1965	Secondary Teacher Education 1976
No. of schools	48,954	45,879	897	16,233	25	26
Enrolments	6,878,336	7,533,739	422,328	3,490,398	12,605	16,153
No. of teachers	192,545	248,354	20948	175,348	808	1,442

Source: Data from Hebei Provincial Educational Commission.

1984 decision to formulate major laws on education in the next several years. The definitive reformulation of the earlier decrees came in 1985 with the "Decision of the Reform of the Education System." This has been the guiding document of reform for all levels of education during the reform and opening-up years. The major aims of the reform were to bring about the four modernizations: to increase state funding for education; to ensure that the education system shall supply a sufficient number of highly qualified personnel; to institute a nine-year compulsory education policy; to expand the system of technical and vocational education; and to give provisions for reform of higher education, for example, to change the system of job-assignments to graduates, and to grant the colleges and universities more decision-making powers.

The law on nine-year compulsory education, which took effect on July 1, 1986, established requirements and deadlines for attaining universal education tailored to local conditions. It also guaranteed school-age children the right to receive education. People's Congresses at various local levels were, within certain guidelines and according to local conditions, to decide the steps, methods, and deadlines for implementing nine-year compulsory education in accordance with the guidelines formulated by the central authorities. The program sought to bring rural areas into line with their urban counterparts.

Academically, the goals of the reform were to enhance and universalize primary and junior secondary education; to increase the number of schools and qualified teachers; and to develop vocational and technical education.

One of the first changes in higher education after the end of the Cultural Revolution was the restoration of the national unified university entrance exams in 1977.

As mentioned earlier key schools were shut down during the Cultural Revolution, reappeared in the late 1970s, and, in the early 1980s, became an integral part of the effort to revive the lapsed education system. Because educational resources were scarce, selected institutions —usually those with records of past educational accomplishment— were given priority in the assignment of teachers, equipment, and funds. They also were allowed to recruit the best students for special training to compete for admission to top schools at the next level. Key schools constituted only a small percentage of all primary and secondary schools and funneled the best students into the best secondary schools, largely on the basis of entrance scores. In 1980, the greatest resources started to be allocated to the key schools that would produce the greatest number of college entrants (In 8, 2001).

Popularized Primary Education

Hebei Province declared popularized primary education in 1985 after an assessment and approval by an evaluation group organized by the provincial government.

Under the education reform after the Cultural Revolution, a major concern in Hebei was that scarce resources should be conserved. This meant that secondary education should not be blindly pursued while quality primary education was still developing. Money, teaching staff, and materials should not be diverted away from primary schools as it was during the Cultural Revolution.

In fact, in Hebei Province, the enrolment in primary education was higher in 1975 than in 1993, and the enrolment rate in 1975 had reached the same level of 1985 (see Table 3.8 and Figure 3.14).

Although by 1975 the percentage of students enrolled in primary schools was high in Hebei Province, schools reported high dropout rates and regional and gender enrolment gaps. Most enrollees were concentrated in the cities and more rural girls than boys dropped out of school. Furthermore, the conditions of the physical facilities of primary schools were extremely poor in rural areas, especially in the mountainous areas,

Table 3.8
Enrolment and Enrolment Rate of Primary Education in Hebei (1975–2004)

Year	1975	1980	1985	1993	1999	2004
Net enrolment rate (percent)	97.7	97.0	97.7	98.50	99.94	99.8
Enrolments (in millions)	7.87	7.35	6.01	7.76	7.76	0.715
No. of school-aged children (in millions)	6.05	5.94	4.57	7.36	7.36	0.5147
Percent of school-aged children	12.32	11.50	8.24	11.63	11.63	7.56

Source: Data from Hebei Provincial Educational Commission and *Hebei Economic Yearbook 2005.*

Figure 3.14
Enrolment of Primary Schools in Hebei

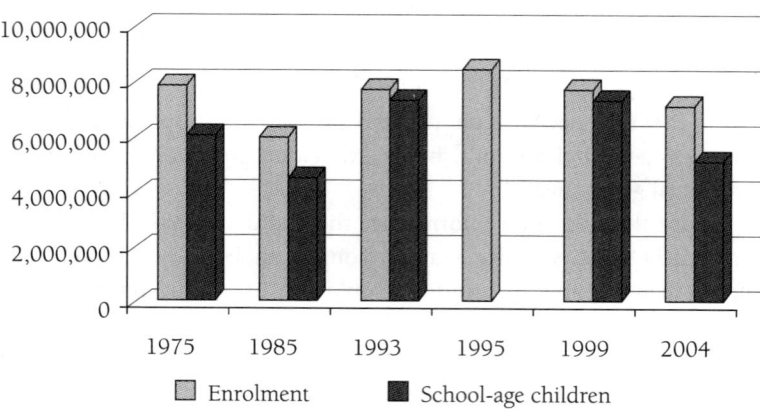

Source: Data from Hebei Provincial Educational Commission and *Hebei Economic Yearbook 2005*.

and many pupils were attending classes in dangerous schoolhouses and were being taught by untrained *minban* teachers. (Salary is not given by government, it is given by the people from the district. Here, *minban* is the same pronunciation of the Chinese character.)

The impact of regularization on the schools led to a number of them closing and merging, especially among the primary schools. In the rural sparsely populated areas, this resulted in declining enrolments. Rural parents were generally well aware that their children had limited opportunities to further their education. Some parents saw little use in having their children attend even primary school, especially after the establishment of the agricultural responsibility system, which began in 1978. This provided for a remuneration system based on output in rural areas, and so for some families, education for their children, as opposed to working at home, was not always the most advantageous choice. Under that system, parents preferred that their children work to increase family income—and withdrew them from school—for both long and short periods of time.

Under the Law on Nine-Year Compulsory Education, primary schools were to be tuition-free and reasonably located for the convenience of children attending them, thus pupils could attend primary schools in their villages. Parents paid a small fee per term for books and other expenses such as water, electricity, transportation, food, and heating during winter. Previously, fees were not considered a deterrent to school attendance in Hebei, although some poor parents felt even these minor costs were more than they could afford.

An Illustrative Case: Sanhe County: The First County to Attain Popular Primary Education (PPE) in Hebei Province (FI4, 2002)

Sanhe County is the first county that reached the standard of PPE set by the Ministry of Education in Hebei Province. As early as 1983, the primary education completion rate of children aged between 12 and 15 had reached 96.1 percent. In that year, 98.6 percent of school-aged children in the county had already participated in primary education and the dropout rate had decreased to 1.7 percent.

In fact, parents valued their children's schooling in Hebei Province. The main problem of the achievement of the goal of PPE was the lack of resources to improve the physical conditions of primary education in rural areas.

Sanhe County had 351 primary schools in its 395 villages in 1983. In 1978, the county had approximately only 40,000 sets of desks and chairs, and 37 percent of the schoolhouses were identified as dangerous. In order to improve the facilities of primary education, the county closed 36 senior secondary schools and 54 junior secondary schools, and focused their financial resources to develop primary education. The schoolhouses and furniture of the closed secondary schools were used to improve the physical condition of primary education. From 1973 to 1983, the county invested 700,000 Yuan to improve the physical condition of primary education. During these 5 years, they built 7 new key schools and repaired about 1,000 classrooms in 70 poverty-stricken production brigades. The county government announced that it would provide financial assistance to the communes or production brigades, respectively 50 Yuan for repairing each classroom and 100 Yuan for building each new classroom. The county government had bought 350 m² of wood to make desks and chairs for primary schools. Furthermore, enthusiasms of the communities had been aroused, and villagers of the county were active in supporting schools of their village to improve physical conditions since the government transferred the rural primary schools to local administration.

Popular Nine-Year Compulsory Education

The May 1985 National Conference on Education recognized five fundamental areas for reform to be discussed in connection with implementing the party Central Committee's "Draft Decision on Reforming the Education System." The reforms were intended to make the county-level government responsible for developing "basic education" and to systematically implement a nine-year compulsory education program.

The Law on Nine-Year Compulsory Education, which took effect from July 1, 1986, established requirements and deadlines for attaining universal education tailored to local conditions and guaranteed school-age children the right to basic education. People's Congresses at various local levels were, within certain guidelines and according to local conditions, to decide the steps, methods, and deadlines for implementing nine-year compulsory education in accordance with the guidelines formulated by the central authorities. The program sought to bring rural areas, which had four to six years of compulsory schooling, into line with their urban counterparts.

Chinese secondary schools are divided into junior and senior levels. Junior secondary schools offer a three-year course of study, which students began at 12 years of age. Senior secondary schools offer three-year course, which students begin at the age of 15. Since 1986, popular junior secondary education has become part of the nine-year compulsory education.

In 1985, the 5,423 junior secondary schools in Hebei enrolled 2,123,700 students. In 1999, the junior secondary school enrolment in Hebei increased to 3,844,400 students, while the number of junior secondary schools lessened to 4,272. The promotion of primary school graduates to junior secondary schools increased from 70.1 percent in 1985 to 98 percent in 1999 (see Table 3.9).

The desire to consolidate existing schools and to improve the quality of key secondary schools was, however, under the educational reform, more important than expanding enrolment (see Figure 3.15).

In 1999, another 23 county-level units in Hebei Province have passed assessment and approval of popular nine-year compulsory education by an evaluation group. This group is made up of the CCP Provincial Committee, the provincial government, the provincial People's Congress, and the provincial political consultative conference, and the number of county-level units approved by the provincial evaluation groups as counties that reached the goal of popular nine-year compulsory

Table 3.9
Basic Data on Junior Secondary Education in Hebei by Area (1985–2008)

Year		No. of Schools	Enrolment	Primary School Graduates Promotion Rate (percent)	No. of Teachers	Pupil–Teacher Ratio
1985	General	5,423	2,123,700	70.1	128,700	16,5
	Vocational		44,179		2,534	17.43
1999	General	4,272	3,844,400	98	199,300	19.29
	Vocational		49,950		2,748	18.18
2008	General	3,484	800,800	99.84	193,632	14.16
	Vocational	311	185,100		15,731	138.58

Source: Data from Hebei Provincial Educational Commission.

Figure 3.15

Enrolment of General Junior Secondary Schools in Hebei

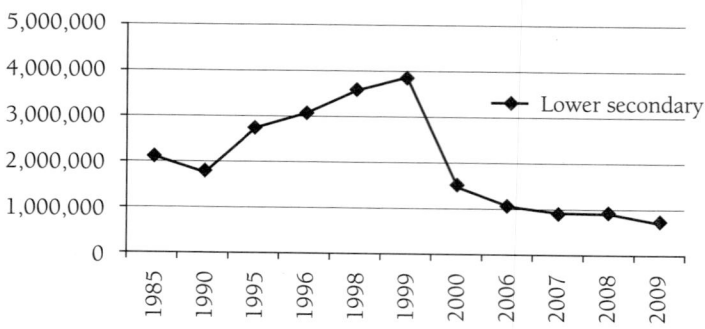

Source: Data from Hebei Provincial Educational Commission.

education totaled to 170, which covered 98.5 percent of the province's total population.

Technical and Vocational Education in Rural Hebei

China's secondary schools are evaluated in terms of their success in sending graduates on for higher education, although efforts persist to educate young people to take a place in society as highly valued and skilled members of the work force.

In rural China, a senior-secondary-school graduate is considered as an educated person, although secondary schools are viewed as a training ground for colleges and universities. And, while secondary students are offered the prospect of higher education, they are also confronted with the fact that university admission is limited. Beginning in 1976, the "unitary" approach of the Cultural Revolution was criticized for its ignorance of the need for two kinds of graduates: those with an academic college preparatory education and those with specialized technical and vocational education. After 1978, the serious problem appeared that a large number of secondary school graduates returned back to their villages without any practical skills and they could not meet the needs of rural economic development. To develop secondary vocational education then became the main task (FI7, 2003).

With the renewed emphasis on technical training, technical schools reopened, and many general secondary schools were converted into vocational schools. The secondary school consisted of four types:

1. Key general secondary schools,
2. Non-key general secondary schools,

3. Specialized technical secondary schools, and
4. Vocational secondary schools.

In spite of the need for technically trained manpower for the economic reconstruction of China, the acceptance of technical and vocational secondary schools was slow, at least initially. The perception lingered that these educational streams were only for those not able to pass in the traditional stream to climb the social ladder through higher education. In 1978, enrolment in technical and vocational programs in Hebei comprised only 1.29 percent of the total enrolment. However, in 1998, of the junior secondary graduates who continued their schooling, 9.69 percent preferred specialized education in technical or vocational schools, the highest proportion recorded, whereas 90.31 percent entered general secondary schools (see Table 3.10.)

However, after 1978, vocational education, including agricultural technical education, recovered and developed rapidly in rural Hebei. In addition to educational authorities, other sectors also felt the need for technically trained manpower and developed their own technical and vocational schools or short-term training programs. In this context, in one county there could exist more than one independent technical and vocational schools or short-term training courses. Statistics shows that in 2008 there were altogether 267 small vocational secondary schools in Hebei, an average of 2 in each of the 142 counties and county-level cities of the province that enrolled 976,500 students (see Figure 3.16).

In order to reduce the management costs, Hebei Province created the county-level comprehensive vocational training center (VTC). By the

Table 3.10
TVE Enrolments in Secondary Education in Hebei (1978–2009)

Year	1978	1995	1998	1999	2009
Total TVE enrolment		3,345,100	4,607,400	4,939,700	201,940
Proportion of TVE enrolment (percent)	1.29	7.26	9.69	9.42	14.60

Source: Data from Hebei Provincial Educational Commission.

Figure 3.16

Enrolment of Technical and Vocational Secondary Schools in Hebei

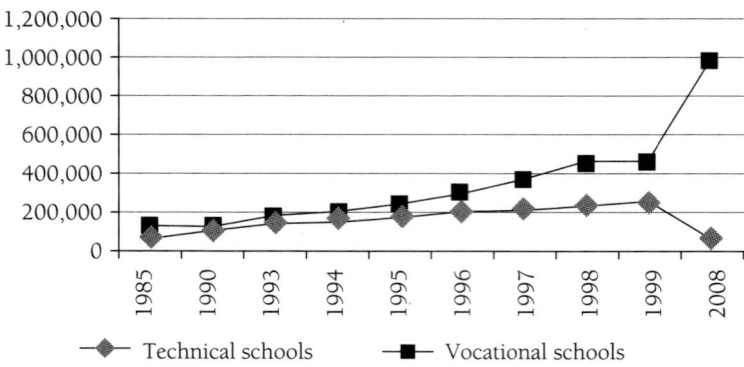

Source: Data from Hebei Provincial Educational Commission.

end of 1995, each of the 139 county-level units of the province had established a VTC through merging the vocational secondary schools and short-term courses run by different government departments or quasi-government organization.

A county VTC is administrated by a board chaired by the head of the county with the directors of the related departments of the county. The county VTCs offer one-year, two-year, three-year, and four-year, as well as short-term training programs for junior secondary graduates. Apart from fostering trained technical workers for the society, VTCs also conduct applied scientific research, provide technology extension, conduct production demonstration and business workshops, and conduct research in vocational education.

An Illustrative Case: Beihaoqian Agricultural Junior Secondary School (FI3, 2002)

An agricultural junior secondary school in Hebei practices a system of 3+1, which after finishing the three-year courses of general junior secondary education as part of the nine-year compulsory education adds a one-year course on practical agricultural technology. It is open to any primary graduate who is willing to study in it. The agricultural junior secondary school offers 3,060 class-hours of ordinary courses, which are the same as general junior secondary schools, and 1,020 class-hours of specialized agricultural courses in the four years. Beihaoqian Agricultural

Junior secondary School, located in the rural town in the hilly plateau, was established in 1985 on the basis of a general junior secondary school and under the administration of the township government. It has 21 teachers and administrative staff, 4 regular classes and a short-term class enrolled 224 students. The school has an agricultural farm of 3.73 ha, a grassy slope of 6.67 ha, a wood farm of 1,987 young trees, and a small animal farm.

According to local needs, the school provides courses in three specializations: agronomy, animal grazing, and fruit planting. The main special courses are an introduction to farming, soil and fertilizer, vegetable planting, sheep-raising. With the help of experts of the county Science and Technology Commission, teachers of the school developed local textbooks such as practical technologies of animal raising and crop planting.

In the school, experimentations on the school farms are also part of the curriculum.

A girl who graduated from the school, started raising rabbits and earned 2,000 Yuan in the first year after graduation through selling 110 rabbit feeders. She also provided 150 rabbit feeders to her fellow villagers and transferred scientific rabbit raising methods and knowledge on rabbit disease prevention to them. With her help, some other households had specialized in rabbit raising, and it developed into a small industry in the whole village. Later, she was employed as a rabbit-raising technician by the school and assigned to teach in training classes on rabbit-raising technology.

Technical Education for Peasants

Literacy education and technical training courses for peasants were other components of basic education in rural Hebei. In 2010, the literacy rate of Hebei had reached 97.39 percent and registered a record higher than that of the state standard for basic elimination of illiteracy (see Table 3.11).

In order to sustain the results of literacy education, the Provincial Education Commission of Hebei prepared a series of applied technical courses like horticulture, fruit tree management, livestock feeding, fishing, and so on, suitable for rural Hebei, and established spare-time primary and secondary school for peasants

In 1980, the priority of education for farmers in Hebei started to transfer from running literacy classes to establishment of part-time primary

Table 3.11
Literacy Rate of Hebei

Year	1982	1991	1995	1999	2010
Literacy rate (percent)	87.6	95.2	92.25	95.87	97.39

Source: Data from Hebei Provincial Educational Commission.

Table 3.12
Basic Data on Peasant Education in late 1990s

Year		1995	1998	1999
Technical secondary school for farmers	No. of schools	18	13	14
	Enrolments	4,263	4,621	4,208
	Teachers	357	285	269
Part-time secondary school for farmers	No. of schools	121	1,492	43
	Enrolments	6,982	150,769	4,213
	Teachers	215	702	361
Part-time primary school for farmers	No. of schools	2,798	6,672	2,232
	Enrolments	87,094	138,752	89,532
	Teachers	820	1,127	548
Part-time agricultural technical school for farmers	No. of schools	34,892	70,781	48,048
	Enrolments	3,030,700	5,410,400	3,995,100
	Teachers	8,660	17,505	9,726

Source: Data from Hebei Provincial Educational Commission.

and secondary schools and farming technical schools. In 1999, the 43 part-time secondary schools for farmers in Hebei enrolled 4,213 students, the 2,232 part-time primary schools for farmers enrolled 89,532 students, and the 48,048 part-time agricultural technical schools for farmers enrolled 3,995,100 (see Table 3.12).

In Table 3.12, almost all data in 1999 is less than that in 1998. There are two reasons, one is because that from 1998 the adjustment for the schools started, many schools have been merged, another is because the farmers income from agriculture declined; therefore, farmers paid less attention to learn something to improve the agricultural productivity because of even increasing more productivity, the income from agriculture were still very few due to the unreasonable price of agricultural products.

Conclusion

This chapter has discussed and analyzed the data, as well as built up the knowledge base for Hebei Province, China, studied in terms of its educational, political, economic, social, geographical, philosophical, and other background. The next chapter (Chapter 4) will further focus on AUH and its rural development activities, and the following two chapters (Chapters 5 and 6) will represent the same knowledge base for NT and CDU, all of those will fit the two stages of description and interpretation necessary for the later juxtaposition and comparison as determined by the research methodology used in this research.

4

Agricultural University of Hebei and Its Rural Development Case Studies

Introduction

This chapter discusses the role of higher education in the development of rural areas within the Chinese context. One agricultural university in Hebei Province (i.e., AUH) had been involved in the advocacy of an education model "Combining Theory with Practice." This model has proven to be suitable in the context of rural China for its fast effects and substantial outcomes. The education model is the key strategy in implementing "The TMM: A Road to Prosperity," which is a development model that integrates education, science, and technology in reconstructing the rural economy and reclaiming the ecological environment of the mountainous areas (Zhou, Li, and Zhang, 1990, p. 56). The university which has been mandated by the state government to serve rural areas has responsibility over farmers living in these remote, impoverished regions. The university responded to this challenge when university professors had decided to move out from their university offices and embrace a new principle (to combine theory with practice) and provide leadership in the reformation of the vast rural areas of the province, especially that of the Taihang mountainous regions. In a period of 20 years (1979–99) of extension work in these regions, farmers have not only become self-sufficient but have also ended their life of poverty and have become more progressive (INRULED, 2000, p. 13).

A Brief Review of AUH

AUH, founded in 1902, is one of the oldest higher agricultural educational institutions in China and now is one of the key universities of Hebei Province. There are 25 colleges with 51 undergraduate programs, 24 programs for master's degree, and 4 for doctorate degree (http://www.infoplease.com/ce6/world/A0835976.html, accessed on November 25, 2010). Currently, the university has developed into an institution mainly of agriculture with agronomy, engineering, economics, and management as core specializations.

The main university campus is located in the ancient city of Baoding, Hebei Province, about 138 km south of Beijing. Baoding is known as the "South Gate" of the state capital (Beijing). The university has an area of 617,342 m² with a 300-ha university farm located only 4 km away from the main campus. Its experimental forestry base (with total coverage of 2,700 ha) is located near the Western Tombs of Qing Dynasty in Yixian County. In addition, there are two other campuses, one is situated in Qinhuangdao city with an area of 73,000 m² and the other one is in Dingzhou city, 60 km south of Baoding city. A campus is under construction in the Western part of Baoding city.

The total number of the faculty and staff members is about 2,000, and the total student enrolment is more than 21,000, among whom are about 800 postgraduates for masters and doctorate degrees (University Information Handbook, 2001, p. 3).

Functions

The University has a trilogy of functions which are teaching, research, and extension (AUH, 1998, p. 21).

Teaching

In catering to the needs of the 21st century, an integrated discipline system centering on agriculture has been formed which includes forestry, engineering, and economics. This enables the integration and combination of different specialization. The coverage of the various fields of study is thus expanded. In order to develop qualified scientists and technicians equipped with both solid agricultural science foundation and professional knowledge, competitive and encouraging systems have been introduced. The integration of teaching, research, and extension has been implemented with an emphasis on the development of skills. This policy was completely sanctioned by the Chinese Ministry of Education.

Research

Scientific research is fostered as a general goal of national economic development and reconstruction. Taking the application and development of research as the main aim of economic development, both basic research and hi-tech research are used as the main branches of study. The policy has already achieved some good results. Since 1991, the university has made 225 achievements in research programs of different levels, of which 68 reached the international level. The university has obtained 134 awards above the provincial level, of which 8 were awarded by the state government and 2 were awarded by United Nations Development Programme (UNDP) (http://www.hebau.edu.cn/xuexiaogaikuang/nongdajianjei.html, accessed on March 16, 2010).

Extension Services

Under the Chinese education system, each province has at least one agricultural university. The agricultural universities are mandated by the state to serve the agricultural and rural needs of the province. As such, AUH has to play a major role in servicing the rural areas, and it has a duty to the farmers in Hebei Province.

AUH has a very strong background of extension services that span over 30 years (from 1979) when it started the implementation of the Taihang Mountain project that led to the development of the TMM: A Road to Prosperity (Zhou et al., 1990). This development model was based on the experiences generated while doing an integrated research and extension program in developing the Taihang mountainous regions. The main aim of the integrated research program was to reclaim the ecological balance of these vast mountainous regions while at the same time economically reconstructing the area. The case study of the TMM development is the main focus of this part of the thesis.

AUH Mission and Vision

Traditionally, this university (AUH) emphasizes the integration of teaching and agricultural production, theory, and practice, so as to train students to become agricultural technicians equipped with sound basic knowledge, high skills, and good practical experiences (*University Information Handbook*, 2001, p. 3).

From this mission statement, AUH has formulated the following guiding principles and ideology (*University Information Handbook*, 2001, p. 5).

The main task of the university is to train qualified personnel. It emphasizes on four main education principles: (1) to carry out national education policies, (2) to serve the development of the rural economy, (3) to put the principle of teaching as priority, and (4) to persist in integrating theory with practice (AUH, 1995, p. 6).

To Carry Out National Education Policies

The university adheres to the policy of carrying out the national education priorities, for example, in 1992, a notice on actively implementing the integration of agriculture, science, and education with the purpose to improve the development of rural economy (AUH, 1997, p. 59). University personnel are inculcated to become academically competent as well as morally upright and physically healthy public contributors of the country. The students are cultivated to love the peasants, uphold agriculture, and to serve rural development.

To Serve the Development of the Rural Economy

AUH advocates a policy that education should fit in with the needs of economic development. According to this policy the key to the development of agriculture and rural economy lies in the training of qualified personnel, making scientific research, and extending and transferring these technologies back to the farm. The vital objective of education is concerned with training qualified personnel to become practitioners (Wang, 2000). The university curriculum contains both basic theory and practice. AUH scientific research applies and exploits the achievements from the scientific research. These research outcomes are then utilized to extend and transfer knowledge and technologies from AUH according to the conditions of different localities.

To Put the Principles of Teaching as a Priority

In the process of promoting rural development, AUH decisively manages the interrelationship between teaching, scientific research, and production. Teaching is seen as a priority. The system of teaching, combining theory with practice, had been strengthened and improved. Qualified practitioners have been recruited who will promote rural development (Wang, 2000, p. 6).

To Persist in Integrating Theory with Practice

The integration of theory with practice will promote the development of both teaching and research. It will advance the teaching level of AUH and stimulate the succession of scientific research. From the development of fundamental theory, study projects are made and applied, leading to new concepts being discovered. Through effective feedback, new concepts are gathered and integrated into the teaching course content. These new concepts are serving as inputs for new scientific research (Wang, 2000, p. 11).

AUH (1997) has always attached importance to scientific research and the extension of science and technology. Concrete outcomes like awards, development of new breeds, strains, and varieties generated from AUH research projects have played significant roles and provided benefits to users. These results are highlighted (http://www.hebau.edu. cn/xuexiaogaikuang/nongdajianjei.html, accessed on March 16, 2010).

The Development of the TMM

In this section, a model developed by AUH in their 20 years of practice in integrating theory with practice—combining extension, research, and teaching as a tool for productivity will be discussed. The focus will be on how the AUH was able to combine educational reforms with economic reforms to achieve rural development.

Taihang Mountain is the largest mountainous area in Hebei, China. Changing its poor underdeveloped status and its ecological environment is of great strategic significance for Hebei Province in order for it to be economically strong. It is also a key area for the improvement of the whole ecological environment of Northeast China (Zhang, 1992, p. 62).

The TMM

TMM is a developmental model. It has been generated from the continuous experience of AUH for more than two decades using an approach of integrating teaching, research, and extension education. The model combining theory with practice is used in the reconstruction of rural areas and the reclamation of the ecological environment of the Taihang mountainous region (Wang, 1998, p. 8).

A Brief Review of the Taihang Mountain Project

In view of the adverse natural environment and many farmers living in poor condition, the exploitation of the Taihang mountainous area is concerned not only with the improvement of ecological environment but also in finding immediate solutions to alleviate farmers from poverty. The experts from AUH conducted investigations and analyzed the characteristics of the mountainous area in 1979. Based on the results of the investigations they made, and with due consideration of society, economy, and natural resources, it was proposed by these experts to adhere to the following development principles: "poverty transformation before mountain transformation, ignorance eradication before poverty eradication and intellectual transformation before poverty transformation" (Zhou et al., 1990, p. 56). It was decided that exploitation should be initiated from available resources using some advanced technological projects. Investment was to be small but with fast effect and great economic benefits, in order to lift farmers from poverty quickly. Comprehensive exploitation was started with an experiment in combining demonstration and extension services, comprehensive development in combination with comprehensive management, exploitation technology in combination with economy, and poverty transformation in combination with ignorance eradication and disaster prevention. Based on the scientific research and analysis, the university's professors and staff have provided very useful and practical advices and suggestions to the farmers, for instance, how to grow strawberries in simple locally made greenhouses, how to better grow persimmon trees in mountainous areas, how to better raise rabbits, how to improve crops production, etc. Thanks to theses services from the university, after the first three years (in 1981), many farmers in the mountainous area (experiment site) had greatly improved their productivity and increased income generation, and also they were freed from poverty and the ecological environment in this area saw some preliminary improvements (Zhang, 1992, p. 125).

Chronology of TMM Development

Introduction

The involvement of AUH in rural development that led to the development of TMM is not made possible by chance. It is emphasized that during this period, the central government mandated every sector of government to be involved in rural development. The public outcry wanted to reform in all aspects: education, economic, etc. In both urban and rural sectors there was a thirst for reformation but the need was more urgent in the rural areas (Zhou et al., 1990, p. 19).

When It Started?

In 1979, Hebei Provincial Government made a decision to develop the Taihang mountainous region (AUH, 1998, p. 6). It established a research project, which was named the "Comprehensive Research and Exploitation of Taihang Mountainous Area in Hebei Province" (Zhou et al., 1990, p. 5). AUH undertook the responsibility for this project and organized professors, teachers, technicians, and hundreds of students in six departments of the university to initiate the comprehensive exploitation of the Taihang mountainous areas.

Initial Achievements of the Experimentation Was Brought in Line with the State Plan

The local government affirmed and set a high value on the exploiting of the model of this mountainous area experiment. Thus, in 1981 a key project called "Research on Exploiting Taihang Mountainous Area in Hebei Province" was established and was brought in line with the state plan. Hebei Provincial Government organized some local institutions and cooperative units to accomplish this key project. Again, AUH bore the technological responsibility. Later the exploitation of the mountainous area was expanded further in 24 counties of Hebei Province. The research project was satisfactorily accomplished as a result of the coordination of numerous scientists, technicians in Hebei Province, and farmers in the Taihang mountainous area. There were 40 achievements made in this project among which 16 achievements come up to advanced world standards. (Achievements are further discussed in succeeding sections; AUH, 1998, p. 325.)

Initial Expansion of the TMM

After 1985, AUH further developed the TMM expanding it from the research base (experiment site) where it was created. TMM was extended to the plain and Heilonggang area in Haihe River Valley. The methods of exploitation were adjusted from single technology, for example, apple trees pruning, into the development of multiple branches of science turning the place into a comprehensive economic area. Then, the mainstay industries were established according to different local conditions. The development was extended from the scientific research as the main point to the teaching in combination with scientific research and production. The achievements of the exploitation were introduced and integrated both into higher education and teaching methods reforms. This initiative led to the further development of both teaching and scientific research levels of the AUH. At the same time new achievements and new technologies of the university were also used back at the extension bases. More social and economic benefits were achieved in the Taihang mountainous area. While the technical team focused on education, combining science and technology with economy, economic, and social benefits took effect jointly with ecological benefits. The development of an explicit holistic new approach using scientific methods in extension education was based on the key principle "combining theory with practice" (AUH, 1996).

The Affirmation of the Development Model by the Central Government

The research project on exploiting the mountainous area (started in 1981) was appraised by the central government in February 1986. The acknowledgment at the national level showed that the approaches, methods, and the measures applied in this research project were commendable and that they had quite an evident effect. The project has shown a new explicit way to manage and exploit mountainous areas, which was acceptable to the central government (Zhou et al., 1990, p. 263).

Further Expansion of the TM Model

Since 1991, experts from AUH have further extended the Taihang Mountain experience to most parts of the plain in Hebei Province. The area covers from the Huanghai to the plateau of Bashang Zhangjiakou district. More than 20 relatively stable three-in-one (teaching, research, and extension combination) extension education bases were set up in different types of ecological regions, including 98 three-in-one work sites for teaching, research, and production. The scientific and technical expansion area covers 78 percent of the whole province (AUH, 1998, p. 161). A lot of activities were done for improving the farmers' scientific and cultural awareness, centering on sustainable development. This is evidenced by the cases in this chapter.

The project content, strategies, approaches, and methods used in the implementation of TMM are discussed in the next section.

The Project Content and the Way of Running It

AUH launched massive income-generating activities (IGA) aimed at opening up of main areas of production and development of mainstay industries at county levels in the Taihang mountainous areas, for example, marmot rabbits raising and related industries became the mainstay in Linzhang county (Case 8 and In 3, 2001). The general objectives of the project were (a) to educate people on the knowledge of science and technology to improve farming practice; to improve productivity to have not only more food for consumption, but also in order to gain more income from the sales of their produce. (b) To develop mainstay industries while simultaneously developing the rural enterprise to absorb the surplus labor force into it. (c) To reverse the condition of the deteriorating natural environment of the Taihang mountainous area that threatened the sustainable development of North China in both the physical and economic environments (Wang, 1999, p. 78).

Specifically, the project aimed at three specific objectives such as:

1.   To make full use of the local natural resources and put them into the market.

2. To introduce and spread practical technologies among the villagers to elevate their income.
3. To bring in technology items which needed fewer funds but were easily done and could get much profit in a short time (Wang, 1999, p. 79).

The Team and Resources

The project was carried primarily by AUH and its task group composed of experts drawn from its faculty and staff. A number of students were also mobilized under their social practice (generally speaking, that means students spend some time in factories, farms, and mines instead of in classrooms to know what the real situation of society is and also to help local people). The project has preferential support from the provincial government of Hebei through the different bureaus and government organizations in the province. Linkages with the local industries, private organizations, NGOs had been pushed through during the course of the project implementation bringing in both personnel and logistics support. Total financial investment amounted to 7.2 million RMB (Zhou et al., 1990, p. 68).

Training Materials and Methods

There was a vast variety of methods and materials used in the implementation of the model. These included bulletin boards, blackboard, poster, broadcasts, technology prescriptions (white paper), slide shows, video tape film showing, scientific advertisements, on the spot live performances using indigenous approaches (e.g., operas), farmers' night schools, winter schools, one-on-one/face-to-face technical consultations, experimental bases and demonstration spots, inputs service stations, science and technology market, model households, and a snowball dissemination effect (In 4, 2001).

Basic Approaches Employed in the Taihang Mountain Project

There were many approaches adopted in the Taihang Mountain project. These were (In 5, 2001; In 6, 2001; AUH, 1996, 1997, 1998):

1. Experts made a comprehensive system analysis, a holistic approach in considering the existing problems and potentialities

of the Taihang mountainous area. Based on the data (as of 1979), although natural resources are rich, the output value is only 0.00012 percent of the total provincial output value. 1978 statistics recorded an average annual income of local residents as low as 50 RMB only (Zhou et al., 1990, p. 68).

2. Setting up of experimental bases. Based on analysis, suitable crops were introduced to enhance productivity. Existing crops with great production potential were also developed. Research projects were implemented and based on the results of the study— scientific-based technologies were derived and extended to the farmers. Experimental bases served as focal point for demonstration spots, which were aimed at disseminating the agricultural technologies, which are coupled with the setting up of service stations (like demonstration stations) for both technology and farm inputs (AUH, 1998, p. 21).

3. Establishing a system of extension and training for agricultural technology and overall quality improvement of the science and technology workforce. This was accomplished by:

   i. The compilation of practical technologies based from actual research done in the site.

   ii. Technological training of farmers to adopt necessary technologies that require less work hands but with high production efficiency.

   iii. Organizing a rational technical delivery service system (i.e., consultative group) under the technology network.

   iv. Advocacy in setting up of various professional technical societies (e.g., Fuji Apple Development Society) to serve in the consultative group.

   v. The establishment of scientific and technology market of AUH (in 1988) linking university with farmers for mutual benefit.

   vi. Selecting poor counties (10 out of 39 in the whole province) to set up experimental villages to implement the "well-to-do-village" project to carry out poverty alleviation strategies.

   vii. Pushing forward the Liao Yuan Plan and promoting rural professional education (hand in hand with Spark Plan, Harvest Plan, etc.) (those plans are national projects that focused on rural education reform, agricultural technology extension, and poverty alleviation).

viii. Training a large group of farmers to become chief members of an extension work force under the Green Certificate Program. (*Note*: The Green Certificate is a national program for farmers conducted by the Chinese Ministry of Agriculture as well as an award given to farmers who have undergone technical training and who have proven potential and capability to extend the acquired practical skills. They must also have some management and supervision capabilities. The farmers who obtain the Green Certificate have many benefits; for example, it is easy for them to have a loan from a bank.)

ix. Setting up Beigu Farming School with assistance from AUH experts trained 16,000 farmers among whom 240 were appointed as farmer technicians by the County government—which led to the farming output in 1993 having raised by 56 percent compared to 1988 (AUH, 1998, p. 223)

4. Mobilizing human resources. A scientific technical extension team was established. AUH was the base unit, vocational students were the means and farmer technician were the main participants. Every year in over 20 departments of AUH about 500 specialists and researchers, over 100 masters degree students and 1,000 bachelors' degree students go down to the rural areas, living and staying in farms and assisting farmers (AUH, 1998, p. 225).

5. Building up/providing a powerful system of leadership (prefecture, county, and university levels) to make up a complete body of policy-makers, administrative support, and implementing units. This was achieved by (In 6, 2001):

   i. Organizing a logistical service system (e.g., distribution centers of production inputs).

   ii. Drawing up an overall plan; making feasible quotas (structural plans of the annual target within five years).

   iii. Evaluating achievements made by research to guide the local production activities.

6. Networking

    i. By joining efforts with local seed company, AUH set up a strong marketing network for Chinese Cabbage seed in Gaoyi County.

    ii. To provide information and training facilities, AUH contributed 6,000 books and reference materials and other teaching equipment to Yongnian County Professional School and to Anping County Beigu Farming School (In 5, 2001).

7. AUH organized farmers in various technical associations to bring into play the initiative of farmers to learn and use science and technology. The following mentions some of these technical associations: Mushroom Association in Tang County, Chicken Association in LaiYuan County, Red Fuji Apple Association in Shunping County, Watermelon Association, Peach Association, Vegetable Association, Maize Association, etc. Farmers' associations opened up market links in China and abroad. Chicken and Vegetable Associations market produce not only in Baoding but also to nearby cities in Tianjin and Beijing. Red Fuji apples gained market access to neighboring countries in Asia (In 5, 2001).

Innovative Approaches Initiated by AUH

Through the development of TMM, AUH accumulated many experiences. Some significant ones are mentioned in the following subsections (In 13, 14, 2002, and AUH, 1998).

Establishing Joint Ventures Using a Contract Package

AUH initiated this new extension service in a range of counties. The prerequisite of this approach was to provide services for farmers. The operation of this particular approach is that the service provider signs a contract package of technical service. The provider charges some fees for the overall service as the resource of technical research. This approach brought a change of delivery pattern from only a special department towards an overall operation pattern by mobilizing many departments like administration, material, supply, financial, and monetary bringing service closer to the needs of rural economic development. AUH and Ding Xing County formally signed the contract for agriculture comprehensive technological package service in January 1989.

On the basis of self-willingness and mutual benefit, AUH established more joint ventures of teaching, scientific research, and social practice bases. By signing contracts with the bases (since the bases were operated independently) AUH gradually changed the extension service mode from totally free to the combination of free and charged services. This new approach is aimed to benefit both sides, the university and local communities to mobilize the initiatives of both providers and recipients and to further enhance the enthusiasm of providers. Through many years of practice, AUH and the local partners of joint ventures have expanded the practice to a bigger scale. The range of services was enlarged from science and engineering to art and soft science, from introduction to the expansion of extension service, from techniques of increasing production to postharvest processing technology, and from economic development to the combination of economic and education reforms. AUH formed such joint ventures with Shunping, Fuping, and Xiongxian Counties from 1982.

Setting up of Science and Technology Consulting Centers (in 1984)

These centers offer various special and technical training courses for rural people. Some courses and consultations are free but some are charged with minimal fees. This income in turn was added to the investment resource of scientific research for purchase of equipment to further enhance research. Moreover, the training courses graduated many groups of skilled labor.

Launching the Social Practice Approach

The social practice approaches initiated by the Chinese Communist Youth League (i.e., Chinese student volunteers with over 200 AUH students in 10 different groups) in 1995 using one-on-one teaching method (i.e., one student assist one farmer, one group assigned in one village). Activities used were broadcasting, blackboard writing, bulletin board announcements, and farmers' night school learning. Course content includes promoting new high-quality products, disseminating new technologies, offering training courses for local agricultural technicians, and delivering technical consultation and on-the-site instructions. A total of 120 social practice teams (generally speaking, that means students spend some time in factories farms and mines instead of in classrooms to know what the real situation of society and also to help local people to do something) were developed.

Deploying Homecoming Practice Teams

This initiative followed on with that of social practice, and it resulted in 1,500 teams being developed by AUH, who submitted 18,000 investigation reports, technical consultations for approximately 40,000 people and 60,000 copies of scientific and technical materials were distributed, 80,000 farmers were provided with technology education classes.

Adopting the Shareholding Cooperation System

The mountain development programs were invested in by the government and implemented by experts and scientists. Farmer beneficiaries received a lot of assistance towards improvement of their life and productivity for free. Continuing this approach may lead to farmers becoming dependent, on assistance, draw money away from scientific research initiatives, then universities, research institutes and government may suffer more economic burdens. In 1996, after the investigation of the aspects of natural production conditions and farmers economic income situation, the AUH Mountain Research Institute (AMRI) contracted with the Provincial Science Commission in "Hebei hilly and dry land agriculture sustainable development comprehensive experiment zone in Taihang Mountain to transform the hills for 50 years through shareholding cooperation with local farmers." The farmers' group (Party A) in Yushanzhuang Village, Gouchang Town in Tangxian County in Baoding area agreed with AMRI task team and both parties signed a contract for transforming and developing the barren hills. Party A provides 134-ha barren hills and an experimental land of 13.3 ha, takes responsibility for engineering, organization, providing labor, and entire production investment including seeds, young plant, fertilizers, and explosives, urging every contracting household to achieve their tasks in time. AMRI as Party B, provides the technologies needed in transforming engineering including engineering design, technical instruction and training, and introducing an appropriate project for Party A, providing all inputs for experimental land production. The profit division is 90 percent shares belonging to Party A, 10 percent shares belonging to Party B. The validation of contract will last 20 years. Party A signed subcontracts with each contracting farmer in the village to ensure the benefit of both sides are protected.

Achievements Made in Hebei Province

There have been a number of achievements through this model. These are (AUH, 1998):

1. *Agricultural production increased and the agricultural structure changed.* Hebei Province is one of the agricultural production regions in China and it has had good harvests for many successive years since 1991. It has reached up to 25 million tons in 1994 and was maximized to 27.8 million tons in 1996 reaching the highest record in the history of the province. The crop harvests have now created the favorable conditions for the livestock breeding industry.

2. The establishment and opening up of main production and development of mainstay industries: for example, Red Fuji Apple in Shunping County, Chinese Cabbage Seed Production in Gaoyi County, Strawberries in Marcheng County, and Marmot Rabbits Production in Linzhang County. Dates, persimmons, walnuts, and maize are among the most products in Taihang mountainous areas.

3. *The development of local enterprises to absorb surplus labor force from the farm.* The contract groups have pushed the development of rural enterprises forward and liberated a productive force. Several forage processing plants have been set up in Ding Xing County; Yan Tai Chicken farm kept on enlarging its operation scale, making, hatching, feeding, slaughtering, processing, and marketing a coordinated process. By the end of 1993, farmers had set up main group enterprises mainly in the fields of car part assembly, hat handicrafts, and chemical production.

4. An investment of 7.2 million RMB in scientific and technological research works got in return an economic benefit of 300 million RMB. (The university and various governments had invested 7.2 million Yuan for projects to develop the Tiahang Mountain area in 1979.) After three years, the benefit was up to 300 million Yuan, which covered the agriculture, horticulture, forest, and livestock in this area (Zhou et al., 1990, p. 68).

5. The average personal income and living standards of inhabitants in rural and urban areas improved. In 1997, the average income

of inhabitants in Hebei Province reached 4,958 RMB and the average income of farmers reached 2,286 RMB, which was an increase of 1.6 and 2.3 times, respectively, compared to 1992.

6. Improved workforce quality through training. The capability of farmers has greatly been improved. For example, farmers became expert seed breeders of Chinese cabbage seeds, the county agricultural and farming machinery schools have trained tens of thousands of people. The technical competence of farmers has increased with quality. Over 2,000 farmers reached the level of agricultural technical workers and over 400 reached the level of assistant technicians and more than 70 farmers to the level of technicians in Ding Xing County alone.

7. *The improvement of the ecological environment.* Now that farmers have been more than self-sufficient in grain, they are willing to develop a diversified economy to reforest the barren hills to improve the ecological environment. In the last 10 years (since late1980s) they have improved the hills of 2,666.7 ha, reforested hillsides of 566.7 ha and planted 96 thousand fruit trees, and established 150 orchards (in Juncheng alone).

8. *People's attitudes change.* A farmer and his wife in Gaoyi County learn to read and write with their son at night to improve his literacy level. The farmer learns in order to read and understand more practical technologies according to him. More and more families are sending their children to study and acquire higher education. Girl children are now given equal chance to study as boy children (In 14, 2002).

Summing up the Achievements Made and Their Implications

Massive social transformations were brought about by initiating economic reforms coupled with educational reforms in the mountainous areas. The former poverty-stricken desolate villages, which were characterized by barren hills, rough treks impassable by any sort of transportation, and low mud huts have now turned into lively progressive villages with accessible road system, green hills crowned with fruit trees, diverse crops growing in the field, tall brick houses with telephones, television sets, and other household equipment that were before considered a luxury item for village people. People's initiatives have also changed not only in money-making pursuits but they now consider their environment also and try or become involved to improve it. Health and sanitation issues that did not appear to bother them much before are becoming a major concern. This is a good indication that rural people are looking forward to a better tomorrow by continuously striving to improve their living conditions. People experienced a liberating force emancipating them not only from the shackles of poverty but also from a feeling of hopelessness and resentment of their situations moving them towards a new freedom of creating their own initiatives in the pursuit for quality life.

Not only the rural families have benefited from the Taihang Mountain project but also the university itself has developed. Below is an account of how the university has continuously prospered while in the process of helping the village prospering.

The University Grows While It Serves Rural Development

Implementing the delivery approach of the three-conjunctions of teaching, scientific research, and production for 20 years, the AUH positively adapted itself to the environment of economic reconstruction, and made its contribution in extension education service to rural development. AUH also developed and upgraded itself. John Burrows has written in his book *University Adult Education of London* (1976) as following:

> University extension has not only served the community by its contribution to intellectual advance and social progress, it has also been of benefit to the universities themselves-by extending their influence and indeed their knowledge of the society that sustains them. More, it has promoted the multiplication of the universities themselves since many of them owe their foundation to the extension movement.

The followings are evidence to further support this opinion.

1. The university gained recognition in December 1996 when AUH was designated as one of the model key universities of Hebei

provincial government. The State government commended AUH in 1986 for the development of TMM (INRULED, 2000, p. 16).

2. The university scale was enlarged. Since 1979, the construction of AUH has been carried out reasonably faster. In 1995, a new AUH emerged from the merging of AUH with the Hebei Forestry College. The size of AUH has been increased in terms of faculties, departments, and number of staff and students. For example, in 1980, AUH operated 7 departments and 10 specialties. As of 2001, AUH covers 16 colleges, 4 departments, and 9 teaching and research facilities. The university offers a total of 29 bachelor degree courses, 24 masters degree courses, and 3 doctorate degree courses. The total number of student enrolment is more than 13,000 with more than 400 postgraduates for master and doctorate degrees. The teachers and other staff members are still in a total of 2000 (the same as that of previous years) (*University Information Handbook*, 2001, p. 82).

3. The specialties were adjusted and developed. In order to fit with the changes of rural industry structure, especially to meet the urgent need of rural commodity production and market economy development, AUH broke out from the traditional concept of teaching of agricultural production when setting up specialization courses. The different fields of specialization were adjusted in accordance with the development of comprehensive agriculture. The social demand changed quickly under the market economy condition. AUH took the attitude of respecting reality and being practical and realistic when putting up fields of specialization. Present needs were considered as well as long-term adaptability so that the university gained a stable and healthy development. Prior to 1980, for example, the course settings of AUH contained some biases: for instance concentrating on production but neglecting postharvest; attaching importance only to crop planting but neglecting product processing; regarding traditional farming techniques but despising new techniques; attaching importance to production but neglecting trade and sale. This approach resulted in course settings being not responsive to the complexity of a changing rural economy and diversity caused by farming modernization (*University Information Handbook*, 2001).

Some Significant Cases

This section will deal with eight cases AUH has conducted during more than two decades of participating in rural development. Two decades is a long period for the university staff to be in rural areas. There would be many stories during that time. As the limitation of both time and space, only a few could be presented here to be shared. The aims are "to identify some features and to show how they affect the implementation of the systems and influence the way an organization (such as a university) functions" (Bell, 1993, p. 11). The cases included:

1. Empowering villages through education, science and technology;
2. Fruit growing;
3. Training the farmers and extension workers;
4. Establishment and development of scientific and technical market in AUH;
5. Sustainable development;
6. Organizing farmers' associations;
7. Group contract;
8. Livestock growth.

From the presentation and analysis of these cases and further data concerning, specific generation about the role of the university in rural reconstruction can be produced.

Empowering Villages Through Education, Science and Technology: The Chaichang Village Case (In 5, 2001, and FI1, 2001)

The first case to be examined concerns empowering villages through education, science, and technology: the Chaichang village case. At the end of 1995, AUH participated in the project of Empowering Villages through Education, Science, and Technology initiated by the Chinese Association of Agriculture. After consulting the Yixian County Government and the Chinese Association of Agriculture, AUH organized experts study visit to Yixian County in cooperation with the Provincial Government of Hebei. As a result of the field investigation Chaichang Village was identified and selected as the first pilot village of the project, where action research and theoretical analyses were to be undertaken in the pilot program.

By 1999, AUH had extended its programs into 9 villages of 3 pilot counties where about 800 AUH professors had offered about 600 various technical courses for about 80,000 trainees, distributed about 60,000 copies of training materials, donated about 10,000 scientific and technical books and periodicals. As a result 100 of the latest techniques out of AUH research projects were introduced to the farmers, which made a great contribution to the economic growth and social progress in the pilot villages and the counties.

The approaches or strategies of implementation that were adopted are many and varied; a few significant ones are highlighted in the following subsections.

Approach 1

An expert from AUH is assigned by AUH to live in the village (year-round) to provide technical service to the farmer residents. The selection is based on the person who has a relevant program and some other specific strategies, for example, spending one year in a rural area will have a free of English exam before to be promoted to a higher academic level.

This is supplemented with a door-to-door campaign plus a cash incentive that is provided to pioneering farmers who attended technical training.

During the period of 1996–99, 50 technical training classes were conducted on forest and fruit trees, animal husbandry, and crop planting. These have benefited 6,000 farmers/times (which means 6,000 farmer once or probably 3,000 twice) and technical key members of more than 100 persons/times. Now, it is common that in each household in the village there is always one person who has mastered at least one or two practical techniques for livelihood. Thirty-six villagers were awarded the Green Certificate. Some women farmers became excellent in growing fruit trees.

Approach 2. Utilizing Existing Resources and Improving Production by Injecting Scientific Management

In this case, two fruit tree types existed in the village but their production was too low. One is the persimmon plantation consisting of 6,000 persimmon trees. It was also found out that this variety of Millstone persimmon was the best kind and suitable under local conditions. The technical experts pruned these trees. But before that, local technicians signed contracts with two model households to identify 10 demonstration trees to be managed scientifically. Harvest production increased. Using these two model households 80 percent of the population in the village began to prune their trees and become technology adopters. In 1998 yield reached a high record of 300,000 kg and in 1999 another 350,000 kg yield was recorded in the village book. Realizing the value of science and technology in agricultural production, farmers began to develop hilly land and grow more fruit trees—while reforesting wild mountains. A single tree of persimmon can produce 250–300 kg and at best times even yielded 500–1,000 kg per tree with scientific management. The issues of marketing were considered especially when persimmon production was reaching this record high. The production of persimmon has become a large-scale industry today and because of the good quality the persimmon from this region demands a good price.

Another fruit tree is the wild apricot where about 6,000 wild trees are found in the mountains. The fruits ripen and drop down with no one noticing them because these fruits are difficult to store and because of the fruit characteristics (it rots quickly after ripening) is not convenient to transport. In 1997, AUH organized students to make systematic investigations of usable resources in this village. Experts from AUH introduced a variety of apricots with thin flesh and big kernels. It was realized that planting new seedlings and for the tree to mature and bear fruit takes time. Therefore, experts introduced at the same time a grafting technique—where the upper part of the wild apricots is cut and a graft of the new variety on the wild ones is undertaken. This new variety taken from Zhangjiakou has proven to be of good health and the kernels have been processed into a famous beverage.

Approach 3. The Introduction of New Varieties and Production Techniques and Other Technologies

Under the assistance of AUH and the Chinese Association of Agriculture, other fruit varieties and grain varieties (i.e., 2 wheat, 4 maize, 22 rice), apple, grapes, Chinese chestnuts, walnuts, lean meat pigs, beef cattle, etc., were also introduced. Up to 95 percent of the village was involved. Other practical techniques included up to 60 different technologies in agricultural production were adopted, for example, combined production technique of increasing the quality of wheat and maize, vegetable planting in greenhouses; cut-flower growing, pruning, and grafting of fruit trees; and the use of methane tanks for power and light. In the

Chaichang Village alone there are 35 established greenhouses for vegetables, flowers, and seedlings.

Approach 4. Protecting Vegetation by Grazing Sheep and Cattle in Fenced Areas

In the past, in this village all the domestic animals (cattle and sheep) strayed in the mountains, so they destroyed the surface vegetation. This practice did not accord with the aim of sustainable development of the mountainous areas. Under the guidance of the technicians, all the domestic animals have been raised and fed in confinement, instead of having them stray freely in mountains. This method is favorable for raising and feeding techniques and also such a practice protects the surface vegetation of the mountainous regions.

The Case of Fruit Growth (In 2, 2001, and In 10, 2002)

The second case concerns fruit growth in Shunping County. Shunping County, one of the poor counties in Hebei Province, lies in the eastside of Taihang Mountain, with total area of 708 km², population of 264,000, and an arable land of 26,800 ha. In 1981, the average income per year per person was only 73 Yuan RMB.

In 1983 based on the investigation of the region by technical experts from AUH, the Red Fuji apple was introduced. Through research and development, over 10 years practical techniques were packaged and extended to farmers. By 1995 nearly 400 ha of Red Fuji apples gave an output of 30 million kg with an output value of more than 100 million RMB. This made Red Fuji apples a major industry of the county.

Mancheng County is an area of strawberry production. It lies in the eastside of the Taihang Mountain. The fertile soil and climate are suitable for growing strawberries. However the yield was low due to poor varieties, mono-planting, and backward farming techniques. There were only 80 ha planted with strawberry in 1980s with the yield of 10 ton per ha. At that time strawberry production was of little importance to the local economy. Based on investigations made by local leaders together with AUH Horticulture Department it was proposed to set-up a base of strawberry production in Mancheng due to the strategic location of the county neighboring major market in cities like Beijing and Tianjin.

Initial improvements gave good results, using the research capability of AUH Department of Horticulture and a variety of strawberries were collected both local and from other countries. From 1981 to 1985, a total of 100 varieties were introduced to the base. Seed plots were set up to provide a stable source of planting materials for successive production seasons. Research continued on increasing yield and producing varieties that were resistant to pests and diseases. Local farmers were highly motivated and took strawberries production seriously. The initial area planted with high-yield varieties increased from 20 ha in 1992 to 1,000 ha by 1997. With more technology support, for example, off-season growing of strawberries in greenhouses, the most recent figure of area planted with strawberries recorded 2,200 ha taking up 50 percent of the total area of Mancheng.

The Case of Training the Farmers and Extension Workers (In 13, 2002, and FI2, 2001)

The third case, professional education, occupies an important position in the national economy of China. There are close links between education and science, and technology and economy, and each can be switched over in the modern work force directly and efficiently. The development of professional education especially in the rural area links that of the rural economy with the improvement of farmers' quality. Since 1987, the university (AUH) actively took part in rural teaching reform and has made contributions to the setting up of county-level vocational and technical education centers working with Hebei provincial and local government. The university has also taken part in the construction of centers for professional education and other activities, for example, holding varieties of training classes and writing a compilation of practical texts. In 1999, the university developed a special course enrolling three groups of farmer students coming from the rural areas; they were chosen and paid for by the provincial government. These farmer students were trained with practical production skills and were required to go back to the farm after their graduation. Now, many graduates gained many achievements in agricultural production. In 1997, the Institute of Adult and Vocational Education of AUH was established to enroll students from the county level vocational and technical education centers. The students were trained here and most of them have been back to their own counties.

In 1995, the university held training classes in Fuping County (an old liberated area but one of the poorest and the most backward) on agricultural technology and management which lasted for one and a half years. Trainees included both students and village leaders. The students were mainly graduates from high schools who returned to the rural area and worked in the county over two years. The town and village leaders, who mainly studied the relative technologies in agriculture, administrative management, basic laws, economic laws, and the management of town and village enterprises, etc., which could help them lead the masses in getting rich. The university also held lectures on special practical technologies. Thirty graduates from the class had been assigned by local governments as the heads of villages after they graduated with college certificates.

The Establishment and Development of the Scientific and Technical Market in the Agricultural University of Hebei (In 6, 2001, and FI5, 2002)

The fourth case to be examined concerns the establishment of the technical market. In 1988, in order to help the development of rural marketing, the AUH set up the General Developing Company of High Technology. Meanwhile, they built an AUH market street—the first market supported by the university in the country. The market served as a bridge between the university and rural areas and it is the extension of the development model of Taihang Mountain. The market brings together up-to-date farming methods and advanced farming technology. The whole business of the market includes the marketing of the varieties of farming materials, transferring technologies and technical consultations, etc.

The guiding principle of the market project is to put social benefits as a priority, as well as to popularize farming technologies, to direct service work, and to spread technical achievements in the rural area. The market has made full use of the resources of the qualified personnel, information and data from AUH. There have been many technical achievements, and the market has now accelerated the transformation of these technical achievements into productive action. The university thus discovered a new way for universities and colleges to open the technical market, and made contribution to the market economic system.

This market has a number of scientific and technical features; for example, business shops selling high yield varieties of over 13 kinds of crops in 1995; 15 kinds of hybrid maize, which used new varieties developed by the university and placed into the market—expanded business up to 24 provinces since 1994. The annual turnover rate is over 100 million RMB.

Another feature of the market refers to the expert clinics which have been set up to give free consultation, and to answer the farmer's questions about crops and livestock pests and diseases. In 1995, the consulting clinics developed over 2,000 pieces of reference data and distributed them. Over 200 thousand farmers, for example, were trained in new technologies used in vegetable growing and management.

The technical market has become the fieldwork site for students in agricultural courses, before graduation. Students gained practical skills and acquired competences.

The market has been an effective way for the university to increase its income. The more prosperous the market, the higher the turnover, so is the profit and income which can replenish the funds of running the university.

The market can link the university with the farm, where problems are used as input for research.

The Case of Sustainable Development (In 1, 2001, In 9, 2002)

The fifth case to be examined concerned Qian Nanyu Village—the typical case that engineering system utilized in "Enriching Mountains and Protecting Plains" has promoted sustainable development of the rural economy.

In 1963, a heavy flood occurred in Xingtai City, Hebei Province. At that time I was teaching in a branch college of the AUH. Wang Fawu, the president, at that time, assigned me to undertake the task to help the local farmers to recover from the flood disaster. I organized a group of teachers and began to make investigations. We decided to have Qian Nanyu Village, Jiang Shui Town, as our pilot site.

The whole village was underwater. The cultivated land was destroyed and the hillsides used for farming were damaged. When we got there, all the villagers had almost no arable land to live on, so they were talking about how to get away from their native village and move to another better place for survival. The terrible situation helped us make up our minds work with the farmers and assist them to change their adverse circumstances.

I majored in water and soil conservation in my college days. After graduation, I taught this specialty in the university. My scientific research projects were also oriented to this direction, so my wish was how to transform the mountains and tame rivers to benefit future generations.

At that time, in Qian Nanyu Village, the situation was very serious, and the people were living a hard life with the per capita annual income of only 50 Yuan RMB per person. The reality there and the villagers' hard life made us even more willing to live together with them and helped them to get out of poverty quickly using our scientific technologies.

Based on the real situation, we thought that the most important thing was how to solve the shortage of food and housing first. So, together with the villagers, we transferred soil to flooded land to reclaim enough basic farmland. When the villagers had food to eat, we organized them to transform mountains and tame rivers scientifically step by step. During the process of implementing the project, we applied ecological economic theory to the construction in mountainous areas, which means to rationally combine ecological economy with transforming mountains and taming rivers. Our task was to change the ecological environment, to solve the problems of flood, drought, poor soil, and erosion which affected the development of mountainous areas and to gain greatest economic benefits was the principle used to implement the project in Qian Nanyu Village. It was this motivation that encouraged us to live and work in the village for many years, and now great changes have taken place there.

In 1999, there lived 375 households in this village, with the population of 1,297 people, among which there were 762 laborers. The cultivated land had been developed to 821 ha, the average cultivated land was 0.63 ha per person. The hilly land was 553 ha, and the average area was 0.4 ha per person. The team clearly realized that not only the limited cultivated land but also the hilly land should be made available for full use to help the local farmers improve their income. After the thorough investigation and research, we developed the technologies of "Enriching Mountains and Protecting Plains," which aims at guaranteeing the lower reaches to be safe when flood happens; accumulating enough water with little rainfall; changing the thin soil layer into the thick soil layer of 1 m to prevent soil erosion. In this way, half of the hilly land became arable.

The procedure was as described below.

First, we made a general investigation to each piece of hilly land in this village, then worked out an overall plan, adopting the orientation of terraces to transform the hilly land. We have been working in the village for more than 30 years. What we did there has thoroughly changed the poor villager with a hard life, bad condition, and intensive physical labor, and with the annual average income of only 50 Yuan RMB per person to a better living place than before. The details of this are that now in this village, the total grain yield is about 550 tons. The villagers have enjoyed the free supply of grain. The total amount of dry and fresh fruits has been increased to 820 tons, among which 320 tons of apples, 125 tons of Chinese chestnuts, and 1,500 kg of walnuts; the total profit reached 2,200,000 Yuan, and the average of fruits is 1,698 Yuan per person. The net income of rural economy is 37,590,000 Yuan, while the industry income is 33,820,000 Yuan (2,800 Yuan per person), the total output value of forestry is more than 2 million Yuan. In Qian Nanyu Village, there are 10 big ravines, among which fruit trees grow everywhere, the winding mountain roads have connected 10 ravines, so it is convenient to go to each of them. To drive through all the orchards takes more than one hour. So one can imagine how many orchards and how many fruits there are. The mountains have become green, and the farmers live a better life. The mountainous village has thoroughly changed into a new one and has become one of the richest villages in the Taihang mountainous region. This village is also named as the "Greenest Place in Taihang Mountainous Areas" by ecological authorities, for its forest cover rate has reached 90.7 percent. As one of the best 500 demonstration sites of environmental protection throughout the world, the village gained the nominee prize awarded by UNEP (United Nations Environment Programme) in 1995.

The economy in the village has developed, and the living standard has been improved. This was part of evaluations to be awarded by UNDP. The villagers' mental attitude has improved, no more fighting and gambling. It has become a famous spiritually civilized village in the county.

In 1998, Mr Song Jian, the state secretary for Scientific & Technology Affair of People's Republic of China, interviewed me and took pictures together with me. After acquainting himself with the details, he stressed that the experiences in Qian Nanyu Village should be popularized widely to benefit the mountainous areas with similar conditions. Now the experiences accumulated in Qian Nanyu Village have been popularized and applied to 65,000 ha of farmland in Hebei Province, which brought about economic benefits of 650 million Yuan in recent years.

The Case of Organizing Farmers' Associations (In 4, 2001 and In 12, 2002)

The sixth case to be examined concerns the organization of farmers' associations. Why the farmers in this mountainous region lived in poverty owed much to their low educational level, and backward farming practices. They could neither apply scientific techniques in their agricultural production, nor could they make full use of the rich geographical resources in the mountains. The principle that the AUH adhered to when it first initiated the development of the Taihang mountainous region was "to deal with ignorance before developing mountains." It is a common understanding in these areas and it is UNESCO's experience in developing countries that only when farmers' cultural knowledge level and the ability to apply science and technology have been improved and their enthusiasm to learn science and technology has been stimulated can the farmers possess the ability to develop themselves soundly. AUH helped to establish various farmers' technical associations based on the farmer paying more attention to outcomes, the real situation of resources, communication, history, geography, human culture, and so forth in different regions. AUH followed such a concrete way, namely, "Experts take the leading position; local government coordinates for them; model households are selected as the core units; farmers are encouraged to join the associations; and agricultural practical techniques are developed first" (personal observation). Thus experts of agricultural technology train members of the associations with advanced practical techniques in agricultural production. They stimulate farmers' enthusiasm for learning science and technology. They also apply this learning through the model households. It appears that the farmers like to learn knowledge actively instead of being educated passively (personal observation).

Based on this principle, the AUH has successfully organized more than 10 farmers' technical associations, which have brought about social and economic benefits. For instance, the farmers have proposed to establish a "Mushroom Association" in Tangxian County, "Chicken Association" in LaiYuan County, "Red Fuji Apple Development Association," "Watermelon Association" and "Peach Association" in Shunping County.

Similar organizations of apple, vegetable, and maize have also been set up in Wuyi and Zanhuang Counties.

The establishment of these associations has played a very important role in promoting the development of the rural economy. It also has been really helpful for farmers to study science and technologies and apply them in an active manner. The establishment of these associations has enhanced extension work and scientific research to improve the farmers' awareness of science and technologies. The Red Fuji Apple Development Association of Beicheng Town in Shunping County was set up by Huangpu Zhongsi, an associate professor in the Horticulture Department from AUH in November 1990. This association has brought about grand economic profits and social benefits. The association not only provides technical training courses, but also instructional services as well as farm inputs and marketing of products. The association is mainly composed of model households while Huangpu Zhongsi acts as the technical consultant. Every year, Professor Huangpu goes to rural areas to hold technical training classes, which will last 1–3 days according to farmers' practical needs in production. The following training programs are included: the management of orchards, the prevention and control of apple trees' diseases and elimination of pests, quantity of watering and fertilizer, storage and postharvest handling of fruits, management of seedlings, even the establishment of orchards, and so on. Each time over 1,000 farmers are trained. Up to now, the association has 10,000 membership, having Shunping County as the center, inclusive of 110 villages in 6 counties nearby. The area of orchards has also been up to more than 2,000 ha compared to a few orchards area that existed for demonstration previously.

The association has the function of integrating scientific research, extension, production, and marketing. Some of the scientific research projects by colleges and universities are carried out in these farmers' orchards. Even a joint "Sino-Japanese Friendly Orchard" by Japanese Huajia Association was set up in one of the association members' orchards. The projects that were implemented there have a close relationship with local production and have the function of demonstration to farmers. For example, scientific researchers have found the latest technique of bag covering for apples, including the best covering bag and the best time for covering, which has shown farmers a new way of improving the quality of apples. This set of new techniques was spread and applied at almost the same time when it was developed. There are some other techniques such as early harvest, quality cultivation and growing, storage and fresh keeping that were also spread and applied, which has greatly shortened the period of transformation of the achievements.

Under the guidance of the associations, orchards are managed scientifically and systematically. The apple output produced by the association members has been up to 37,500 kg per ha, with the total output value of 30 million kg. The yield and quality of apples in Shunping County has exceeded that of Japan, the original place of Red Fuji. The Red Fuji apple production in Shunping County has become a primary industry in the Taihang mountainous region in Hebei Province.

The Case of the Group Contract (In 14, 2002 and Fl6, 2004)

The seventh case to be examined concerns the technical contract group of synthetic agriculture in Ding Xing County.

In Hebei Province, the plain area is more developed than the mountainous area. The application and dissemination rate of science and technology is quicker due to two main factors influencing its acceptability by the farmers. These are high yield per unit area and favorable natural conditions. The continuous development of the agricultural production is affected not only by one or more technical measures but also by a series of factors occurring before sowing, crops growing, and the management before the harvest. So as is in the case of Ding Xing County, agricultural production fluctuated. In order to change the situation and provide the means to coordinate the development of science and technology and link it with economic growth, in 1989, AUH organized the technical contract group of Synthetic Agriculture in Ding Xing County to advance the local economic development. The main measures taken have been described in the following subsections.

A Powerful System of Leadership Was Developed

In Ding Xing County, a leading group was set up. It consisted of the leaders from the prefecture, the county, and the university. The vice-commissioner of Baoding prefecture worked as the chief of the group, two vice-presidents from the University as assistant directors, and relevant departments and the staff from the scientific research section of the university as group members. The staff, sent by the university as vice governor of the county in charge of scientific research, was in charge of a coordinating group. The group was composed of the personnel from the prefecture, from the county, and the university. In the meanwhile, many leading groups and offices were organized in each town. Five teachers from the university were sent to the key towns as vice-directors for the scientific research, and similarly as the case for each key village. In this way a complete system of science and administration was formed, which could effectively link with every department concerned. In the system, the overall plans were made, the concrete activities arranged, the technical measures carried out, the contradictions settled, and the materials provided. Thus, a complete body of policy-makers, administrative support, and implementing units was formed.

Organize the Rational Technical Delivery Service System

The university established a consultative group with professors and specialists majoring in agronomy, plant protection, fruit trees, vegetables, veterinary, hydraulics, agricultural machinery, and foodstuff. Thirteen technical contract groups were organized by the university and local governments throughout the county concerning wheat, maize, cotton, peanut, vegetables, plant protection, fruit trees, lean meat pigs, Xialuo-lai sheep, beef cattle, agricultural machinery, irrigation and drainage, and foodstuff. The technical groups consisted of 50 university teachers who were stationed at the test spots and 100 technical leaders from different departments, and about 1,000 village technicians. An agricultural extension network was formed covering the whole county. This network proved to be effective in transferring technologies from the university to the villages. The technicians were key players in this transformation.

The contract group established a close connection between responsibility, rights and benefits, combined production and supply with marketing, took administrative management, scientific technology and supply as a whole, especially with science and technology as its guide. The single department, different individuals and individual technical contracts were combined into a group composed of many departments of different levels and fields. These offered a series of services for farmers during the entire production period.

The group paid special attention to the three key links during the course of performing the contract. First, they defined the contract quotas and made concrete plans. Each contract group's overall macroscopic plan was drawn up by the professors from the university and leaders and technicians of townships. These were then approved by the administrative levels and implemented by the town technicians and farmers. Second, they found the key issues in one area and solved technical problems as they occurred together. Then, they promoted the work in all areas by drawing experiences gained at the testing points.

Since the local farmers did not have enough courage to take risk using the new production technologies, because of fear of failure, the university tried to demonstrate to them the basic points, and then spread the experience from test spots to nearby areas. According to the real situation of wheat production in Ding Xing County, the wheat professors from the university, through the demonstration spots, adopted the "5 s" measures. This means, spreading pest-resistant fine-quality seeds, spreading the technique of applying fertilizer separately both in spring and summer, spreading the application of the compound fertilizer, spreading the dense sowing, and spreading the planting using the new sowing method with less, but fine-quality seeds. The maize contract group centered on practicing "1 f and 2 i," using fine-quality seeds, increasing close planting, and increasing the adoption of chemical fertilizer. In this way, the farmers in Jiu-han village realized the target of 1,000-kg maize per mu (Chinese unit, 1 ha equal to 15 mu) in the same year.

The professors majoring in cotton from the university put forward five technical measures to improve cotton output, namely, use improved varieties, dense planting, scientific pruning, two times fertilizer application during cotton ball growing, and chemical control of cotton pests. At the same time, great efforts were made to disseminate the technology of plastic film covering. In 1991, Dong Hai-bao town realized the target of 50 kg of ginned cotton per mu. They then disseminated this achievement by using the mass media, for example, radio broadcasting, TV spots to supplement the traditional training classes. The county set up demonstration areas on the spot and promoted their work in the whole county by drawing experience gained from the test spots. To make the scientific and technical achievements widely known to the local farmers' 100 mu of test fields, 1,000 mu of demonstration fields, 10,000 mu of spreading fields were established by the county.

Organize Logistical Service System

The effect of science and technology relies on the supply of materials and logistical service. When AUH and local government had organized the leading group chosen by both condition and willingness, a service network with the departments of finance, banking, supply and marketing, materials and oil supply, and town service stations was formed. Technicians can therefore spread technologies among farmers with enough materials needed. This process is physician's "writing out a prescription, and then making it up."

Establish the System of Extension and Training for the Agricultural Technology and Overall Quality Improvement of Science and Technology Workforce

On the basis of a group contract, AUH, making use of its advantages, has combined agricultural production with science and education and enforced the training for agricultural and scientific technology. To carry out the "Liao Yuan plan" as put forward by the National Education Committee, the university extended education outside the university by training the staff from the middle schools and the vocational school as active technology carriers to spread the agricultural technologies in rural areas. An overall plan was drawn to train the young farmers to master the necessary skills in agricultural production. Now a great number of technicians and managing persons are constantly emerging. In the meanwhile, the experts of science and technology from the university, combining with the local technicians, renew their knowledge to meet the need of the rapid changes in the countryside, through further education.

Draw an Overall Plan, Make Feasible Quotas

The aim of setting up the agricultural and technical contract in a county as a unit is to build the county into a demonstration area composed of all kinds of productive factors, combined with the full utilization of all resources, a rational structure of industries with low cost and high efficiency. All above needs an overall scientific and a well-drawn plan, upon which, feasible quotas can be drawn up. Based on investigation of the area, the university built an overall model system which included agriculture, forestry, animal husbandry, fishery, and rural enterprise, and a four-model subsystems: crops planting, fruit processing, animal husbandry, and rural enterprise. A structural plan of the annual targets within five years for overall agriculture was worked out. The plan not only reflects the production quotas, but also took into consideration the economy, society, and the environment to bring the rural economy into a good shape.

Take Part in the Macroscopic Policy-Making and Microscopic Guidance of the County Government

Through the overall group contract and participation in the policy-making, the university was able to send their vice-magistrates and vice-directors who were in charge of science and technology to guide the local production work. With the help of the comrades in charge of finance and banking, clear directions on the distribution of the quotas and the agricultural investment were made. In order to bear the risks, AUH team cooperated with the Science Committee to make concrete plans and workout the measures for the scientific items and to evaluate the achievements made by research. By means of full participation in every respect, the university aided the development of agricultural production and economic development, both in macroscopic policy-making or microcosmic guiding.

Since the group contract was carried out by the university with the cooperation of the departments of the county, the rural economy has made great progress, as evidenced below.

Firstly, the output of food, cotton, edible oil, meat, fruit, and vegetables has increased sharply. In 1993, the total output value of agricultural products reached RMB 450 million, 25 percent higher than that of 1988. The total yield of food reaching 305,923 tons had increased by 60 percent. Cotton yielded 862.5 kg per ha, increasing by 116 percent. Latest peanut produce is 3,015 kg per ha and total yield gets 1,5637.8 tons, an increase of 90 percent and 149 percent, respectively. The output of fruit increased from two million kg to 6.5 million kg, 10,000 kg increase each year. The annual average income got to RMB 740 Yuan per person, RMB 196 Yuan more than that of 1988, with an average increase of RMB 39 Yuan each year.

Secondly, the capability of farmers has been greatly improved. The county broadcasting and television stations held more than 80 technical lectures, reaching about one million in attendance. The county vocational training schools provide the villagers with 150 specialists each year. The county agricultural and farming machinery schools have trained about 10 thousand people within five years. Every year, the university accommodates more than 10 graduates from senior middle schools for direct and professional training. The technical contingent of farmers has increased with quality. Over 2,000 farmers reached the level of agricultural technical workers. Over 400 reached the level of assistant technicians. More than 70 farmers got to the level of technicians and one senior technician. One farmer technician is in charge of 33.3 ha of farmland, an assistant technician for 133.3 ha, and a technician for 667 ha.

Thirdly, the contract group has pushed the development of rural enterprises forward and liberated the productive force. They have set up several forage-processing plants; the annual consumption of grain is 20 million kg. Yan-tai town's chicken farm, for example, kept enlarging its operation scale, making hatching, feeding, slaughtering, processing, and marketing a coordinated process. The annual number of processed chicken reach by 500 thousand heads. These were delivered to the markets of Beijing, Tianjin, etc. By the end of 1993, the farmers set up five main group enterprises, mainly in the fields of car parts assembly, hat manufacture, handicrafts, and chemical production. There are also many other privately run enterprises that emerged.

The Cases of Livestock Growth (In 3, 2001 and In 11, 2002)

Case 1. The Farmers in Linzhang County Raised Marmot Rabbits

The conditions of Linzhang County

Linzhang County lies in the southern part of Hebei Province, the eastside of Taihang Mountain, and it is named after the Linzhang River, with the total area of 744 km², with the farmland of 750 thousand mu. There are 14 towns with 425 administrative villages and the population of 550 thousand, among which 520 thousand are farmers. It mainly relies on the farming, and few industries with poor financial sources. In 1994, the average annual income per person was RMB 894 Yuan and the number of marmots rabbits raised was less than 1,000. There were no scientific bases of raising them and there was no breeding farm for marmots rabbit then.

The village, Dongcianfangbiao, has 468 families with population of 2,033; total farmland area of 1,680 mu and only 0.7 mu per person to cultivate. The majority of the farmers in the village took crops planting as their life means. There were also 112 vehicles in the village, mainly used for transportation. Before the end of 1994, none of the families raised marmot rabbits and only 3 families fed 15 heads by March 1995.

The marmot rabbits mainly are grass-eating little animals; farmers can gain high-benefit with little investment in a short time. The rabbit skin can be used as raw materials of fur coat making and the rabbit meat is high protein–low fat food. Realizing the potential of raising marmot rabbits, farmers started raising marmot rabbits in large scales. In the later part of 1995, the number of marmot rabbits increased rapidly. The village and the county were well known as the marmot rabbit demonstration bases only within three years. During this period, the professor from the university, Zhang Bao-qing, often went from one village to another, to spread the techniques among the local people. He held lectures, conducted training classes, and gave advice and guidance to the local farmers such as raising management, the housing management and location of marmot rabbit's sheds, the choice of marmot rabbits breeds and breeding techniques, and the prevention and cure of the diseases. He also wrote the technical articles and book on how to raise marmot rabbits. Many farmer technicians have now been trained and are greatly involved in promoting the development of the marmot rabbit production.

The profits gained from the rabbit raising

Because of the cooperation with scientific and technical workers and the enthusiasm of the farmers who are raising rabbits, the teachers from the university were more encouraged to promote continuously the development of the marmot rabbits. More farm families have become rich through rabbit raising which pushed forward the local economy resulting in high benefits gained by the society. The annual average net income per person in Dongqianfangbiao village was only RMB 890 Yuan in 1994. Then after March 1995, the only 3 families raising rabbits soared up to 96 families with the income RMB 40 thousand Yuan the same year. More and more families were attracted to join marmot rabbit raising. By the end of 1996, there were 263 families with RMB 680 thousand Yuan income. In 1997, the total income of RMB 1.16 million took up 80 percent of the total village income with 378 families involved. There were also 3 family rabbit farms. The number of the rabbits of the whole village was 16 thousand heads, and 46 thousand of rabbits were sold with the total income up to RMB 2.20 million Yuan. The average personal income of RMB 1,100 Yuan was achieved in 1997. Under the leadership of the Branch of Party Committee of the village, mostly all village folks young and old, both men and women are preoccupied and are quite busy with rabbit raising. They have no more time to waste and less time to gambling. For example, women are busy with rabbit raising and men are transporting supplies and marketing produce. There are still some who are idle and play mahjong but they are becoming fewer and less than before. A popular jingle, *Push over the mahjong table, set up the rabbit sheds* is embraced by the villagers. Rabbit raising has enabled the farmers to live well.

Case 2. A Rich Man through Raising Rabbits

Mr Zhao Erzeng is a local farmer of a village keeping a five-member family including his mother, his wife, and two sons. He likes raising rabbits in his spare time. His 10 heads of rabbits bred 100 little lovely and cute rabbits. But soon after an overnight, all rabbits died. This shocked him and his family very much. Later Mr Zhao heard that there was an expert in rabbit raising named Prof. Gu Zilin in AUH. One day in 1987, Mr Zhao, with an education level of primary school, spent a whole day on writing a letter to Prof. Gu for help. It surprised him that he received Prof. Gu's reply in just five days. In the reply letter, Prof. Gu told Mr Zhao some problems, which should be paid special attention to when raising rabbits and encouraged Mr Zhao to write him often. After several rounds of letter communication, Mr Zhao was benefitted quite a lot and wanted to see Prof. Gu himself. One morning, Mr Zhao got into Prof. Gu's office with a wondering mood. Prof. Gu explained to Mr Zhao many techniques in raising rabbits. He told Mr Zhao that if one wants to become wealthy, he must be literate in both science and techniques. Prof. Gu gave Mr Zhao some reading materials about raising rabbits and a list of books, newspapers, and periodicals to be bought and read. Mr Zhao worked according to Prof. Gu's advice. Mr Zhao applied the knowledge learned from the books, practicing the technology learned, and got a satisfying effect. This in turn increased Mr Zhao's interest in learning more about science and technology. Mr Zhao's attitudes towards learning changed from passive to active. Mr Zhao and his wife learnt basic literacy from their sons at night. Mr Zhao noted down the question and problems for future consultation with Prof. Gu. Gradually, Mr Zhao improved his techniques of raising rabbits, and told the other villagers with pride that he had a brother professor in AUH.

With the instruction and help of Prof. Gu, Mr Zhao's family changed greatly. Firstly, Mr Zhao's rabbit farm grew larger and larger. Mr Zhao has got 300–500 heads of breeding rabbits so far and been keeping

1,000–3,000 heads of rabbits in his farm for rabbit meat. The old farm was small and not enough for the large-scale production, so Mr Zhao recently built a new large rabbit farm at the village. Secondly, Mr Zhao's income increased continuously these years, from RMB 2,000–3,000 Yuan in 1987 to RMB 5,000–6,000 Yuan in 1988. Since 1988, Mr Zhao's income went up to RMB 20,000–30,000 Yuan or even more for a year. Mr Zhao had paid off all his debts and bought some new furniture and facilities. The Zhao's family has become the rich family in the village.

Conclusion

In this chapter, AUH's activities in connection with agricultural extension and rural development have been discussed. At these points of view, a further looking for cases which might support the general needs and preferences of rural farmers have been developed. Following this, case studies where agricultural extension education and training were used are presented. It is believed that from each of these case studies, we can both learn from the mistakes and be inspired by the results.

5
Educational Development in Northern Territory, Australia

Introduction

The previous two chapters discussed educational development in rural Hebei, China, and a case study of AUH, China. In this chapter and the following one, I am going to talk about educational development in the NT, Australia, and a case study on CDU concerning participation in rural activities, before carrying out the comparative study on both universities. I think this is the best way to understand the educational and social issues, the role of universities in rural development, as well as the requirement of comparative study methodology for description and interpretation.

The initiative for this chapter (Chapter 5) and following chapter (Chapter 6) is to build up the knowledge base about the NT and CDU. During the time I did this, I found that due to the different population sizes and situations between Hebei, China, and the NT, Australia, and the limited information, as well as the limited information available from CDU, the knowledge base is not as vast as Hebei, China, and AUH. This is so for the following reasons: (a) In China, the government pushes the efforts to universities to carry out rural development programs, in the NT, it is a free enterprise government; therefore, the situation is different; (b) AUH has a long involvement with rural services and development programs, CDU has had a short period dealing with rural and remote services; therefore, less data and information from CDU is available; (c) In the NT, governments and other organizations or agents as well as the university are responsible for the service and extension programs in rural, remote, and Indigenous areas, but in China, in that specific period, most programs are carried out by universities. There is imbalance in the amount of content collected regarding the NT and CDU compared to that available about China and AUH. Nevertheless there is sufficient data available from this research to make the necessary comparisons.

The Social and Economic Context

Geographic Context and the Administrative Structure

Geographical Context

The NT is the third largest territory/state of Australia, the total land area is 1,347,525 km² and it occupies 17.5 percent of the total territory of Australia. Surrounding areas include Queensland in the east, West Australia in the west, South Australia in the south, and the northern part is bordered by the Timor Sea, the Arafura Sea, and the Gulf of Carpentaria. The NT is between longitude 129° E. ~141° E. and latitude 11° S. ~26° S, with desert in the south and seas to the north. Darwin is the capital of the Territory. In the north are lowlands, in the southeast are low plains sloping toward the Lake Eyre depression, and in the southwest are the MacDonnell Ranges. The main rivers are the Victoria, Daly, Adelaide, and Roper, all of which drain into the northern seas (http://www.infoplease.com/ce6/world/A0835976.html, accessed on November 25, 2010).

The NT actually has two climates: tropical warmth of the Top End and the crisp desert air of Central Australia. The Top End has two seasons: tropical winter or dry season from May 1 to October 31. The minimum average temperature is 21.7°C, and the maximum average temperature is 31.6°C. This is the time of year with blue sky, warm dry days, low humidity, and cool nights. Another season is tropical summer or wet season from November 1 to April 30. The minimum average temperature is 24.6°C, and the maximum average temperature is 32.2°C. During this season, the Top End has lightning shows, sun showers, warm weather, and high humidity. For the climate of Central Australia there are four seasons: summer, winter, spring, and autumn, but for much of the surrounds of Darwin there are two seasons–the wet and the dry (http://www.ntholidays.com/nt_travinfo.asp?iTravInfoID=4&strPage= Climate, accessed on November 12, 2010).

The northern part of the Territory receives seasonal rains which are heavy and reliable near the coast, but increasingly unreliable at greater distances from it. Much of the Territory is unsuitable for any form of agriculture (Dick, 1986, p. 13).

Administrative Structure

Australia's governance is based on six separate states (New South Wales, Queensland. South Australia, Tasmania, Victoria, and Western Australia) and two territories (The Australian Capital Territory and the NT) (Anderson, 1991). Unlike other states and territory, "On 1 July 1978 the Northern Territory (NT) became Australia's first self-governing territory. In institutional terms, this was a watershed in NT political history" (Loveday and Wade-Marshall, 1985).

The NT has three different administrative levels: Federal Government, Territory Government, and local government. And the local government is divided into city councils, township councils, and village councils.

Administrative Units of Northern Territory

There are six municipal councils in the NT, namely, Darwin City Council, Palmerston City Council, Litchfield Shire Council, Katherine Town Council, Tennant Creek Town Council, and Alice Springs Town Council, as well as 32 community government councils; 29 incorporated associations and 1 special purpose town (Figure 5.1; http://www.lgant.nt. gov.au/loc_gov_info/Lgovnt/municipals.htm, accessed on November 11, 2010).

Figure 5.1
Administrative Units of the Northern Territory

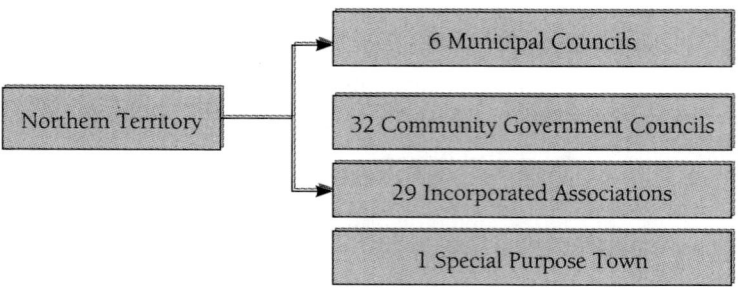

Source: http://www.lgant.nt.gov.au/loc_gov_info/Lgovnt/municipals.htm
(accessed on November 11, 2010).

Economic Development

Present Economic Position of the NT in Australia

The Territory's economic development has been accelerating in recent years. Gold is worked to a small extent; uranium, bauxite, manganese, iron, lead, and zinc deposits are increasingly exploited. Stockbreeding, encouraged by government development projects, is the major rural activity. There is very little farming of national significance in the Territory. However, there are continual experimentations with rice, cotton, and the other grains. Peanuts, pearl shell, and trepang are the principal exports (http://www.infoplease.com/ce6/world/A0835976.html, accessed on November 25, 2010). And "The dominant rural industry of the Northern Territory has been cattle, and it still accounts for most of the Territory's agricultural earnings" (Dick, 1986, p. 1).

1986 is a very old statistic probably 1984 almost 30 years old. You need to upgrade this and maybe talk about the political situation 2011 where the prime minister stopped exports of this beef to Indonesia because of their cattle slaughtering methods. What is the effect on the NT beef industry?

The NT's economy relies on its abundance of natural resources, importance to national defence, relatively large public sector and a short distance to Asia. With the completion of the railway, it is expected that GSP will increase. According to the data of 2010, the territory gross state product (GSP) was valued at around $16.3 billion, which accounts for approximately 1.23 percent of national GDP.

Table 5.1

Industry as a Percentage of GSP/GDP (2010–11) in the NT and Australia

	Unit percentage:	
	NT	*Australia*
Agriculture, forestry, & fishing	3.3	2.4
Mining	17.4	7.2
Manufacture	8.4	8.2
Electricity, gas, & water	1.1	2.2
Construction	10.7	7.7
Wholesale trade	1.4	4.2
Retail trade	3.3	4.5
Accommodation & food services	1.9	2.3
Transport, postal, & warehousing	4.4	5.1
Information media & telecommunication	1.5	3.2
Finance & insurance	4.1	9.7
Rental, hiring, & real estate services	2.5	2.0
Public administration & safety	9.3	4.9
Education & training	3.5	4.5
Health care & social assistance	5.8	5.6
Arts & recreation	1.0	0.8
Professional, scientific, & technical services	3.0	6.6
Ownership of dwellings	9.2	8.0
Administrative & support services	1.6	2.4

Source: ABS No.3210.0

The figures in Table 5.1 show the industrial share of GSP/GDP (2010–11) in NT and Australia.

Economic Development After Self-Government

The NT was formally granted self-government on July 1, 1978, with its capital at Darwin. From that time on, the NT has experienced rapid economic growth. GSP in 2001 has reached $7.45 billion, which is 10 times more than that in 1978. The economic growth includes mining and tourism industries and an important location for the Australia defence forces. Currently, there is a big upgrade of US defence force personnel being housed in the NT. And there is also a potential development of rural and manufacturing industries. But there are economic development constraints in some areas, for example, a relatively narrow industry base, deficiencies in transport links, and isolation from major Australia centers of population and economic markets. However, on March 13, 2002, one important event happened when the Alice Springs to Darwin rail project was realized. Australia Southern Railroad delivered its first locomotive for use between NT and other states. This was an historic milestone for the transportation and for the future development of the NT (Figure 5.2; http://www.railpage.com.au/modules.php?name=News&file=article& sid=54, accessed on August 10, 2010). The figures in Table 5.2 show the economic development from 1978 to 2001 in the NT.

Figure 5.2
Gross State Product of NT at Current Prices ($ Billion)

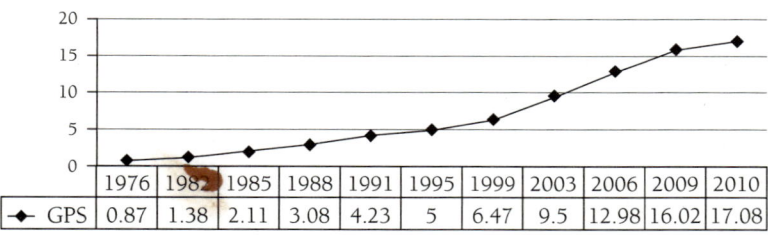

	1976	1982	1985	1988	1991	1995	1999	2003	2006	2009	2010
GPS	0.87	1.38	2.11	3.08	4.23	5	6.47	9.5	12.98	16.02	17.08

Source: ABS No. 5220.0.

Table 5.2
NT Gross State Product at Current Prices (1979–2010) (Unit: $ Billion)

Year	GSP	Year	GSP	Year	GSP	Year	GSP
1979	0.87	1987	2.99	1995	5.00	2003	9.500
1980	1.00	1988	3.08	1996	5.45	2004	10.539
1981	1.22	1989	3.42	1997	5.71	2005	11.758
1982	1.38	1990	3.94	1998	6.01	2006	12.983
1983	1.61	1991	4.23	1999	6.47	2007	14.979
1984	1.91	1992	4.46	2000	7.03	2008	16.420
1985	2.11	1993	4.57	2001	7.45	2009	16.021
1986	2.57	1994	4.62	2002	9.039	2010	17.082

Source: ABS No. 5220.0

Demographic Context

NT is Australia's smallest populated jurisdiction. It accounts for 17.5 percent of Australia's landmass with just 1.03 percent of the country's population. In 2010, the total population of the NT was 229,700 with a population density of 0.17 person/km². Generally speaking, the urban population is much larger than the rural population, and occupies three quarter of the total population. The figures in Table 5.3 show the population growth from 1986 to 2010 in the NT.

In June 2011, the total population of the Territory was 230,200. Over three quarters of the population live in urban centers. The 2006 Census identified that 22.5 percent of territory's population was born overseas, with many from non-English speaking background. And also the Aboriginal population occupied 30 percent of the NT's population and represented 13.4 percent of total Indigenous population in Australia (Figure 5.3).

Table 5.3
NT Population Growth (1978–2000) (Unit: Thousand)

Year	Population	Year	Population	Year	Population
1986	154	1994	173	2002	199
1987	158	1995	177	2004	206
1988	159	1996	182	2005	211
1989	161	1997	187	2006	215
1990	164	1998	191	2007	221
1991	166	1999	194	2008	226
1992	167	2000	197	2009	229
1993	171	2001	200	2010	230

Source: ABS No. 1304.7.

Figure 5.3
Population Growth of NT

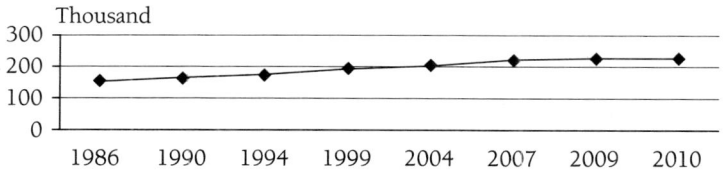

Fertility rate of the NT has changed from high to low. The highest crude birth rate after World War II was 33.7 per 1,000 persons in 1970. In 2009, the crude birth rate was 16.9 per 1,000 persons. According to population projections, the population in NT will go from 250,700 to 380,700 in year 2021 and 506,600 in year 2051(ABS No. 3102.0 and 3222.0).

Since the beginning of the 20th century, with the improvement of living standards in water supplies, sewage systems, food quality, health, education, improvement of social condition, and increase of medical treatment, the NT has experienced a general decline in mortality and an increase in life expectancy. The crude death rate declined from the highest year of World War II 9.4 per 1,000 persons in 1966 to 4.7 per 1,000 in 2000 (ABS No. 3102.0). The average life expectancies of NT were 60.7 years for males and 64.0 years for females in 1971; 68.5 years for males and 74 years for females in 1995; as well as 71.5 years for males and 75.9 years for females in 2010. (The average life expectancies of Australia were 79.5 years for males and 84.0 years for females in 2010, which is among the highest life expectancies in the world.) (ABS No. 3311.7). The main reason why NT's life expectancies were so low is because this number is an average one, which includes a huge percentage of Indigenous population that have very high death rate (Figure 5.4).

The national gender ratio (F:M) is 100:99.2, which is almost balanced. In June 2009, the NT had the highest ratio of males to females of

Figure 5.4
Crude Birth and Death Rate Per 1,000 Mean Population (1945–2000)

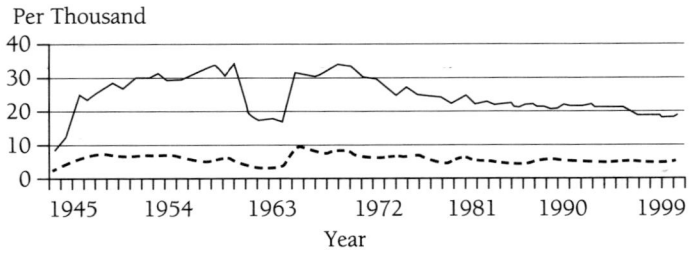

Per Thousand

____ Crude Birth Rate Per 1,000 Mean Population
- - - - Crude Death Rate Per 1,000 Mean Population

Source: ABS No.3102.0.

the states and territories, with an estimated 107.8 males per 100 females (ABS 3235.0).

By the June of 2010, out of a population of 229,700, the population aged less than 15 years old was 23.1 percent of the total; the population of working age group of 15–64 years old was 71.4 percent of the total; and the population aged 65 or over was 5.5 percent of the total. The NT compared to other states and territories has a relatively young population with median age of 31(national median age of 37) (ABS 1304.7).

Educational Development in the Northern Territory After Self-government (After the 1974 Cyclone)

Introduction

On the first of July 1978, the NT became a self-governing territory of Australia. This 30 plus years is a significant period for educational development in various aspects of the NT. My focus in the chapter is only on educational issues after self-government for adult and vocational, Aboriginal, tertiary and TAFE, because lack of statistical data before this.

In the time when the NT became a self-governing territory, there have been some problems in education, such as: high student turnover, high absentee rate, health and nutrition problems, and isolated communities with limited direct experiences. Teachers also faced problems, for example, teachers isolated from the mainstream, difficult living conditions, teachers lacking the opportunities for further education and training, limited teachers and lack of relief staff and so on (Gammon, 1992, p. 31). It is clear that there have been significant improvements to various areas of education in the NT, but there still exist some of the above-mentioned problems, especially in rural, remote areas and Indigenous communities.

The last Report issued by the Federal Department of Education, NT Division was in 1978. In 1979, responsibility for education was transferred from the Federal Government to NT Government. The aim of this Act that transferred the power is: "To encourage more direct involvement by parents in the education of their children. The Act also provides for the creation of local councils for government schools" (NT Department of education Annual Report 1979/1980).

Before 1980, the school design and construction was often a copy of the southern states. After self-governing, it was planned to have smaller, air-conditioned buildings in consideration of local weather conditions, for example, "separated modular units linked by walkways, with provision for natural ventilation and having windows providing natural light and verandahs for shade" (Gammon, 1992, p. 37).

The Annual Report 1982/1983 was a year with another milestone in the NT's educational development. The Annual Report 1982/1983 shows that "The Northern Territory schools' first comprehensive statement of education policies was tabled in the Legislative Assembly by the Minister for Education." (1982/1983 Annual report, p. 5) was achieved. This identified Northern Territory Schools—Direction for the 80s and provided "a framework for the planning and administration of education programs within the Department and in the schools." This concluded many policy directions taken since 1979. "It marked the end of the phase of breaking away from the commonwealth education system which existed in the Territory until mid of 1979" (Gammon, 1992, p. 40).

The Annual Report 1983/1984 mentioned the NT Department of Education's commitment for computer education. From this period, a three-year program had been carried out to ensure that all schools had appropriate computers and relevant equipment as well as providing basic staff training (1983/1984 Annual report).

The Annual Report 1985 gives me a strong feeling of emphasizing many curriculum initiatives and policies. For example, the department provided for effective assessment strategies for literacy, recommending daily physical education, education in early childhood and girls, and also the opportunities for the professional development of teachers.

In 1987, the document—Towards the 90s: Excellence Accountability and Devolution in Education—was issued by NT Department of Education. The Annual Report suggested school improvement plans in all areas, for example, buildings, grounds, human resources, and curriculum.

Accordingly, the NT comes to the 21st century. In the beginning of the new century, the data in 2010 shows that there are 188 schools in the NT including 152 government schools, 36 nongovernment schools, as well as 40 homeland learning centers. The total student enrolment number is 43,494 with 32,978 students enrolled in government schools, and 10,516 students enrolled in nongovernment schools (DET Annual Report, 2010–11, p. 29) (see Figure 5.5).

Figure 5.5

Enrolments by Education Level in NT Government Schools, 2006–10

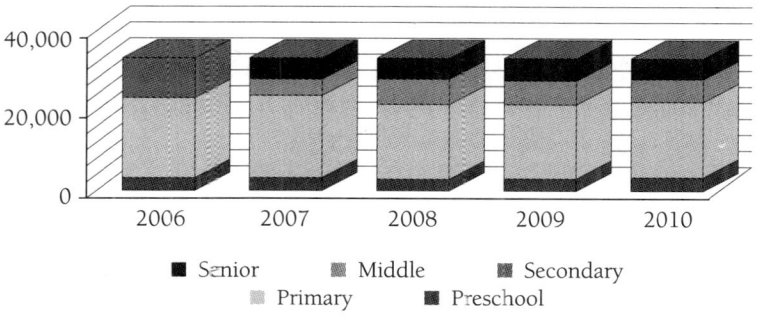

Senior ■   Middle ▩   Secondary ▩
Primary ▩   Preschool ■

Source: DET Annual Report 2010–11, p.33.

As the Figure 5.5 shows, from 2007, the NT moved to a system of middle year's education in which year 10s moved from junior secondary to senior years in 2007, and year 7s moved from primary to middle years in 2008. In other words, primary education now consists of transition to Year 6 while middle years consist of Years 7 to 9 and senior years consist of Years 10 to 12.

Table 5.4 shows that percentage of year 12 students who received NTCE (Northern Territory Certificate for Education). Those who did receive it are eligible to apply for entry to all Australia tertiary institutions.

NT schools have a higher proportion of Indigenous students than any other state or territory. In 2010, Indigenous students covered 40.6 percent of the total student number in NT (DEET Annual Report, 2009–10). From this report, we can also share that NT's Indigenous students have a much lower success rate on national performance benchmarks than non-Indigenous students. Data from this report shows that there is improvement in some areas, for example, in government schools the Indigenous students who achieved Year 3 reading benchmark increased from 29.7 percent in 2001 to 39.9 percent in 2011, achieving Year 3 numeracy benchmark decreased from 65.6 percent in 2001 to 41.0 percent in 2009. Indigenous students' schooling attendance is also another major issue for improving Indigenous education outcomes. In 2009, the Indigenous student attendance rate average was 80.2 percent for primary students enrolled in remote schools. Achievements for Indigenous education are vital factor for the NT rural development, as most Indigenous students live in rural communities.

Table 5.4

Percentage of Year 12 Students Who Received NTCE

	1997	1998	1999	2000	2001	2010
Yr 12 Total	1,129	1,118	1,225	1,126	1,202	1,693
NTCE issued	517	545	579	636	609	1,041
% of NTCE issued Yr 12 Total	46	49	47	56	51	61

Source: Northern Territory Board of Studies Annual Report, 2010, p. 44.

According to 2006 Census, about 44,717 people living in the NT have a language background other than English (DEET Annual Report, 2010–11, p. 23). These live in remote communities, therefore, education and training for those target groups are significant, urgent, and important to build capacity for them to successfully become involved in the surrounding "western society" (DEET Annual Report, 2001–2, p. 88).

Before 1978 self-government of the NT, higher education was weak. In 1989, the NT University was founded on the basis of a merger of the University College and the Darwin Institute of Technology in response to the commonwealth government's announcement of replacing the binary system of higher education, and to meet the increasingly educational needs in the Territory. This university soon became the key role in higher education provisions in the NT and it plays a vitally important role in the NT's economic and social development. It does this through its generation and transmission of knowledge that is essential to the NT's long-term economic growth, social and cultural improvement, and competitiveness. In 2004, Alice Springs' Centralian College merged with NTU to become CDU.

Adult, Technical, and Further Education in the Northern Territory

Education provided as TAFE includes education and training provided by government institutions other than programs of full-time education in pre-primary, primary, or secondary schools; higher education courses; and on-the-job training (selected Northern Territory TAFE Statistics, 1990, 1991). It is a well-established system which covers the whole of Australia. Students enrolled in TAFE study are more flexible, they are full time, or part time.

TAFE in the NT is provided by CDU. The aim of TAFE education is "to meet the technical and further education needs of the NT population who live beyond the city of Darwin" (Northern Territory Department of Education, April 1983, Technical and Further Education Triennial Planning Submission [TAFETPS] for 1985–87, manuscript). This act also requires that educational bodies be concerned with the complex and changing TAFE needs of Aboriginal people in remote communities widely dispersed through the Territory and other urban areas. Attention is also paid to regional TAFE needs and other post-school needs, to the specialized requirements of the Territory's rural industries, to the needs of adult migrants throughout the Territory for whom English is a second language, and to a centralized and local provision of teacher and management education for Aboriginal people (TAFETPS April 1983, for 1985–87, manuscript).

Historically, NT's TAFE education services also are provided through regional centers, which make it possible to extend limited staff resources to remote areas. TAFE has also provided well-organized courses to the various clients to meet the different and specific needs. For example, there is a course called Certificate in Access to Employment and Further Study (CAEFS), which contains name of the course, location of the program, the need for the course, typical positions, mode and duration, course aim, general objectives, methodology and assessment. There are more detailed descriptions for above-mentioned courses, for instance, the methodology used for this course is based on the eight principles of teaching/learning adapted from the Australia Language Levels Guidelines, which are:

1. Students are treated as adult learners with specific needs and interests, therefore, the learning contexts have been taken from real life.

2. Students are provided with opportunities to participate in communicative use of English, in a wide range of activities related to the module being studied.

3. Students are exposed to communicative data that is comprehensible and relevant to their specific needs and interests.

4. Students focus deliberately on various language forms, skills, and strategies in order to support the process of English acquisition, therefore, provision has been made for modeling by the teacher and through oral and written texts.

5. Students are exposed to social cultural data and direct experience of the culture embedded within English language contexts of the local community, therefore, purposeful experiences of the work/ study place have been provided.

6. Students become aware of the appropriate feedback about their progress, through the formative assessment procedures detailed in the document.

Students are provided with opportunities to manage their own learning in western contexts through experiencing a variety of learning activities, and joining in effective group work where they negotiate content and methodology with others. Learning "How-to-learn" strategies are objectives in every module (CAEFS, 1991, pp. 3–6).

Tertiary and TAFE Education

Generally, there are two tertiary education institutions in the NT, which are CDU and Batchelor Institute of Aboriginal and Torres Straight Islander (http://www.nt.alp.org.au/policy/newdirections/ndhedpp.html#key, accessed on August 10, 2010). Both of them were established after NT self-government (after 1978). Both have had significant achievements as well as made a great contribution for educational, cultural, social, and economic development in the NT.

CDU's TAFE Education

Since Chapter 7 will discuss CDU in more detail, my focus here in this section will be in its TAFE education.

As one of Australia dual universities, CDU not only pays attention to higher education, but also has had more emphasis on TAFE education as well, than single sector universities.

In 2002, CDU's TAFE general recurrent profile consisted of approximately 220 courses allocated to 18 industry groups for the purpose of funding (http://www.railpage.com.au/modules.php?name=News&file=article&sid=54, accessed on August 10, 2010).

Table 5.5 shows the data about CDU TAEF students from 1997 to 2001.

From Table 5.5, it is noted that there is not a big number difference between the gender issues of the CDU's TAFE students, but they do have a big difference in attendance, and the most of TAFE students are part

Table 5.5
CDU TAFE Student Data (1997–2001)

Year	1997	1998	1999	2000	2001
Total students	7,574	7,443	8,682	8,793	9,379
Course enrolments	8,691	8,727	10,220	10,681	11,610
Gender					
Female	3,368	3,333	3,905	3,836	4,225
Male	4,206	4,110	4,780	4,957	5,154
Attendance					
Full time	698	499	628	709	703
Part time	7,993	8,228	9,592	9,972	10,907
Age group					
Less than 19	1,208	1,161	1,407	1,564	1,827
20–24	1,430	1,295	1,475	1,355	1,413
25–29	1,149	1,081	1,238	1,175	1,171
30–34	1,002	991	1,142	1,220	1,256
35–39	928	978	1,142	1,220	1,256
40–44	754	738	887	898	982
45–49	530	578	648	660	711
50 or Older	573	559	745	837	882
Unknown	0	62	1	2	23

Source: Statistics Data, Charles Darwin University, Darwin, NT.

time. The interesting thing is age group, there is almost the same number of students from different age groups from less than 19–39 years old, and this seems to me that TAFE education in CDU is much more flexible and suitable for almost all ages of people.

Batchelor Institute of Indigenous Tertiary Education

Introduction: A study of the impact of the universities on rural development in the NT would not be complete without reference to BIITE. Whilst the Institute is not a university at this stage of its development, it has unique characteristics that need to be included in this study for the sake of completeness.

BIITE, formerly known as Batchelor College, began as a small annexe of Kormilda College, then a residential school for Aboriginal students on the outskirts of Darwin, in the mid-1960s, providing short training programs for Aboriginal teacher aides and assistants in community schools. In 1974, the college moved to Batchelor, about 100 km south of Darwin, and has occupied its present site in the township since 1982 (http://www.nt.alp.org.au/policy/newdirections/ndhedpp.html#key, accessed on August 10, 2010).

During 1990, in order to meet the educational needs of Aboriginal people from Central Australia, a second campus was established in Alice Springs. Later in the same year, annexes were opened in Darwin, Nhulunbuy, Katherine, and Tennant Creek (http://www.ntu.edu.au/pvctafero/tafeprofile.html, accessed on July 15, 2010).

After its establishment, BIITE served as an "educational institution for tertiary education of the *Indigenous* people of Australia and the provision of other education training programs and courses, and facilities and resources for research and other related purposes" (BIITE, 2002, p. 2).

From a 1985 enrolment of about 100 students undertaking one-teacher training program, the institute has currently grown to about 3,000 *Indigenous* students with 1,000 enrolled in higher education courses and the other 2,000 in vocational training and education courses (BIITE, 2001) from 143 cultural and geographic communities across Australia; 51 languages are represented (White and Brands, 1999).

Batchelor Institute currently enrolls more *Aboriginal* and Torres Strait Islander students at the higher education level than any other tertiary institution in Australia. The majority of the Institute's students are mature-aged—between 30 and 45 years—while almost 70 percent are women (http://www.batchelor.edu.au/file/webpage/about_history.html, accessed on July 8, 2003). This is likely to change the enrolment statistics, given the recent dismantling of many programs and their transference to CDU and the continual development of providers within the vocational and technical training arena, including corporate industry and mining operators.

The courses at Batchelor are flexible and acceptable to and for *Indigenous* students and communities. They are delivered through mixed modes and with emphasis on community-based study and research, field visit and study, supervised working experience with short-period

Figure 5.6
The Composition of Batchelor Institute

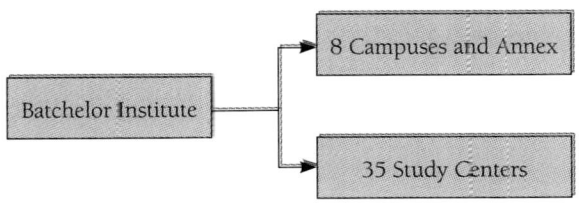

Source: http://www.ntu.edu.au/faculties/site/schools/spi/campusntrc.htm
(accessed on June 3, 2010).

intensive residential workshops in campuses, regional centers, and communities (Donna, 2002, p. 7).

> There are many centres related to Batchelor Institute inside of Northern Territory and interstate regions in Australia. Some centres have their own staff, but most of them have no staff, Batchelor sent staff to these centres to deliver lectures. The assistants or tutors come from local communities, and selected from the former trainees. Most of training is short term courses, 1–2 weeks, and once a term. Normally, most of trainees come back to their own communities after graduation; some gets jobs instead of back to their communities. Batchelor also sent staff to live in communities to do some training courses. (In 24, 2004)

Location and Composition: Batchelor Institute currently comprises campuses, annexes, and study centers in 43 locations throughout the NT and eastern Kimberly region (see Figure 5.6)

Structure: Figure 5.7 shows a structure of Batchelor Institute.

The Batchelor Vision: A unique place of knowledge and skills, where Aboriginal and Torres Strait Islander Australians can undertake journeys of learning for empowerment and advancement while strengthening identity (http://www.batchelor.edu.au/file/webpage/about_locations.html, accessed on July 8, 2010).

The Batchelor Institute has specific responsibility for the provision of teacher education and management training opportunities for *Aboriginal* community people and for those other people who wish to work in communities and are seeking specific orientation for this work. "As well as

Figure 5.7
Structure of Batchelor Institute

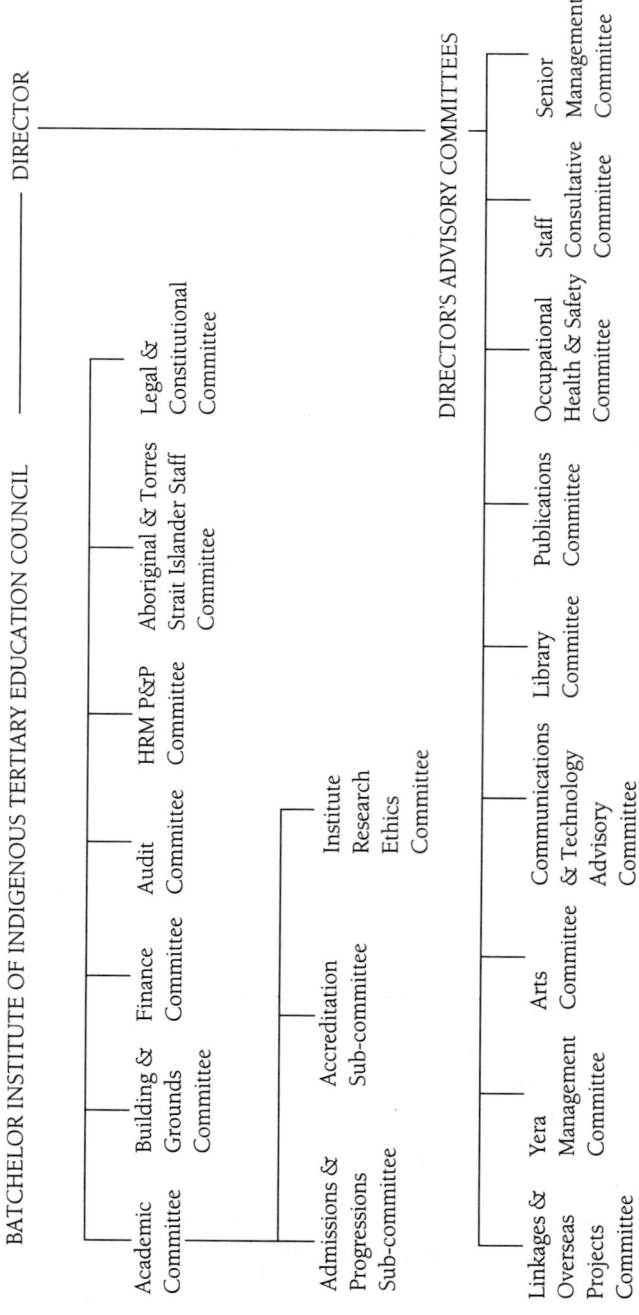

BATCHELOR INSTITUTE OF INDIGENOUS TERTIARY EDUCATION COUNCIL — DIRECTOR

Academic Committee

Building & Grounds Committee

Finance Committee

Audit Committee

HRM P&P Committee

Aboriginal & Torres Strait Islander Staff Committee

Legal & Constitutional Committee

Admissions & Progressions Sub-committee

Accreditation Sub-committee

Institute Research Ethics Committee

DIRECTOR'S ADVISORY COMMITTEES

Linkages & Overseas Projects Committee

Yera Management Committee

Arts Committee

Communications & Technology Advisory Committee

Library Committee

Publications Committee

Occupational Health & Safety Committee

Staff Consultative Committee

Senior Management Committee

Source: http://www.batchelor.edu.au/file/webpage/about_history.html (accessed on July 8, 2010).

the provision of assistant teacher and teacher training for up to three years" (TAFETPS, 1983, p. 37).

Teaching, Learning, and Practice Activities: The teaching and learning activities of Batchelor Institute include different mixed-mode delivering, workshop-based teaching and learning, resource-based teaching and learning, community-based teaching and learning, and so on. Those educational activities could be delivered either in internal or external mode, and have emphasized the practical ways (Donna, 2002).

Most courses of Batchelor Institute are delivered by mixed-mode, which uses community-based learning and intensive workshop learning, and with appropriate guidance to support study and learning (Donna, 2002).

Batchelor Institute also worked closely with the Vocational Training Commission identifying needs in Aboriginal communities and in providing appropriate programs to cover these identified needs. The courses delivered by the Institute are on a regular basis which provides adults with the opportunity of upgrading their basic skill levels in order to proceed to further education and training. For example, the courses on community management type programs have an emphasis on basic business and accounting procedures, management practices for local councils and basic administrative procedures, and so on (TAFETPS, 1983, p. 37). Some programs are designed to develop the skills required for people to manage effectively their own communities. These are short-term specific skill courses. Some of modes are provided onsite in Aboriginal communities, and residential at the Institute, in which students identified from a particular region come together in one of larger communities for the period of the program.

Indigenous Education

The NT is a unique place and includes many culturally and linguistically diverse Indigenous communities with their own knowledge and contributions. An important challenge of the NT is to develop a shared sense of the future with the active participation of all people (DEET Annual Report, 2001–2, Department of Employment, Education and Training [DEET] of Northern territory, p. 115). In order to develop this potential, improving education for Indigenous students in the NT remains a high priority. The NT schools have a higher proportion of Indigenous students than any other state or territory in Australia. In 2005, Indigenous students represented 42 percent of the NT Government's total student population (DEET Annual Report, 2005–6, DEET of Northern territory, p. 25). Therefore, Indigenous education becomes more and more important for NT rural development, and also social, cultural, economic, and sustainable progress.

At the tertiary level, there are two higher education institutions in the NT to provide higher education and TAFE; they are CDU and BIITE.

The objective of Indigenous Education set by Northern Territory DEET is:

> To ensure that all Indigenous students achieve the level of skill, knowledge and understanding necessary for participating in society and undertaking tertiary and vocational education; and that parents, students and the entire community places a high value on education. (IESP, 2000–2004)

Three outcomes for Indigenous education are emphasized: to maximize Indigenous students' attendance and participating in schooling; to meet recognized literacy and numeracy benchmarks and community expectation; and to develop effective system to manage the Indigenous education program (IESP, 2000–2004, p. 2).

One of the key issues to achieve the outcome is to maximize Indigenous students' attendance and participating in schooling is to ensure that Indigenous students go to school regularly, which might be effectively assisted by students' parents, community, and schools themselves. In order to do so, schools and communities should be supported to develop local attendance and enrolment initiatives to keep timely reporting of student attendance and achievement. Families should also participate in their children's education both at school and at home (IESP, 2000–2004, p. 3).

Another key issue for achieving the outcome to maximize Indigenous students' attendance and participating in schooling is to ensure the Indigenous students are healthy enough and that provision of special needs, like disability needs, nutrition, and early language development is a priority (IESP, 2000–2004, p. 4).

Curriculum, teaching and assessment methods, principals, teachers, and school infrastructure need support to ensure Indigenous students achieve satisfactory learning outcomes. And those achievements should be measured and reported to students' parents and government (IESP, 2000–2004, p. 5–6).

Finally, in order to meet the last outcome—to develop effective systems to manage Indigenous education programs, there are still two key issues, how education program is managed with full accountability and how Indigenous family, communities and government can share responsibility for education outcomes (IESP, 2000– 2004, p. 7–8).

Many courses delivered to Aboriginal people are on community training. Their preference is that the services to be provided in their own local communities rather than for the people to travel somewhere for TAFE provision. This on-site delivery includes formal and nonformal education programs determined principally by expressed community needs wherever this is feasible (TAFETPS, 1983, p. 62).

Conclusion

The NT is a unique place with geographic, administrative, social, cultural, educational, and economic diversity different from other states or Territory of Australia. It has a tropical Top End location, self-governing administrative structure, a big group of Indigenous population and a multicultural society. After self-government the Territory's economy has increased rapidly. GSP in 2010 has reached to 16.3 billion dollars, which is 20 times more than that in 1978 (ABS No. 5220.0). Even though the population only occupies 1 percent of total population of Australia, the Territory covers 17.5 percent of Australia' landmass, GSP has reached up to 1.23 percent of the total (ABS No. 1304.7). The Indigenous population occupies 30 percent of NT's total (ABS No. 1304.7).

These have been significant achievements in education since self-government in all sectors (from primary to tertiary, from formal to non-formal, higher education and TAFE) at all levels for Indigenous and non-Indigenous. This has shown up as a great enlargement of school numbers, students' enrolment and attendance number, benchmarks achieved, CDU's establishment and its contribution, Indigenous education, and so on.

After self-government, one of the first initiatives in education was to devise a core curriculum in all subject areas, which made it possible to provide the knowledge and skills necessary for the people to take part in society. Prior to this the NT used other states' curriculum which was typically not suited to the NT as it was only marginally modified.

There were many great changes in NT education after self-government, but the final goals are unchanged, to deliver the quality education in well-resourced schools to meet the various needs of target groups in different communities.

Community involvement is still a priority of school running in the NT. The government provides the facility and staff, and is responsible for operating cost; and the community sets and collects fees and upgrades and maintains curriculum resources (Gammon, 1992, p. 54). In this way, the parents and community representatives are included on schools' activities, such as on Curriculum Advisory Committee and Educational Advisory Council. With the NT Government's policy of devolution, school councils (including parents and teachers) are more responsible for running their community's school, such as deciding the budget and financial priorities, overseeing areas like repairs and maintenance of grounds and buildings, having input into school policies and other decision-making activities (Gammon, 1992, p. 53).

6

Charles Darwin University (CDU) and Its Participation in Rural Development

Introduction

This chapter discusses the Australian context of higher education dealing with rural communities. CDU has been used as the Australian case for this comparative study.

CDU as one of the higher education institutions is the major tertiary education institution in the NT. It provides higher education as well as TAFE-level training. In 2011, there were approximately 7,000 higher education students and 14,000 TAFE students at CDU. Almost all of CDU's higher education courses are offered from its main campus at Casuarina in Darwin. In addition, through amalgamation with Centralian College, CDU offers on-campus degree programs in business and visual arts in Alice Springs. CDU's Palmerston campus, Tiwi (in Darwin) annexe, and regional campuses at Jabiru, Nhulunbuy, and Katherine offer mainly TAFE-level courses. CDU is also the largest research and development provider in the NT, undertaking more than a quarter of all research and development carried out in the NT. CDU has rich experiences in involvement with the activities in rural, remote, and Indigenous areas, which meets the Territory's social, economic, cultural, and environmental needs (http://www.nt.alp.org.au/policy/newdirections/ndhedpp.html#key, accessed on August 10, 2010).

A Brief Review of CDU

CDU is part of the Australian national unified system of higher education. Funded by the Commonwealth of Australia Government, it is a member of the Commonwealth Association of Universities. CDU also receives vocational education and training (VET) funding from the NT Government to provide courses similar to those provided by TAFE colleges in southern Australia (Dondas, 1998). Located at Darwin, capital of Australia's NT and situated in the tropical zone, it is Australia's closest University to Asia. CDU was founded in 1989 on the basis of a merger of the University College of the NT and the Darwin Institute of Technology.

CDU is unique because it is located in the "Top End" of Australia, where it is geographically close to Southeast Asia and its placement is in relation to the northern Australia. These characteristics have provided many opportunities for people to experience northern Australia and to study in a tropical city with a multicultural profile.

CDU's position is also in a developing region amongst countries such as Indonseia and Malalysia which share a similar climate. The aboriginal population represents 30 percent of the NT's population and 13.4 percent of total indigenous population in Australia (ABS No. 3102.0). These factors influence directions in research. Areas of research concentration include Asia Pacific Arts, Energy Studies, Environmental Remote Sensing, Indigenous Resource Management, Tropical Built Environment, Tropical Environmental Science, Tropical Plant Science, Tropical Wetlands Management, South-East Asian Studies including South–Eastern Asian Law, Clinical Nursing Research, Social Research, Studies of Language in Education, and Teaching and Learning in Diverse Educational Contexts. The sector also offers programs in arts, education, business, and science through the faculties in this university, namely, Faculty of Education, Health and Science; Faculty of Indigenous Research and Education;

Faculty of Law, Business and Arts; Faculty of Technology and Industrial Education. Figure 6.1 identifies the current academic structure of the university.

The main campus is in Darwin, 15 km from the Darwin city center. Courses are also provided at Palmerston Campus, 20 km to the south, and at Alice Springs, in the heart of the Australian desert (see Figure 1.3). The NT Rural College at Katherine offers students the opportunity to experience life on a working cattle station. Besides, there are still a few regional study centers located at Jabiru, Nhulunbuy, Katherine, and Tennant Creek, giving the CDU a spread of facilities across the breadth of the Territory and exposure to tropical and desert environments as well as the rich Indigenous culture of Australia. Each regional center acts as CDU's link to regional and remote NT making courses and training accessible to more Territorians.

In line with the national system, the university has identified education, training, and research as its priorities. CDU now has positioned itself as a place for a comprehensive response to the needs of students, industry, and farmers for both VET sector and higher education opportunities with approximately 21,000 students.

Being a dual-sector university (Higher Education and Technical and Further and Adult Education), the only one of its type outside of Victoria, CDU has offerings from the certificate level of TAFE through to doctorates at the postgraduate level. This gives students a number of study pathways. Students can start out in the TAFE sector and progress through to complete their PhD at CDU. The university is formally linked to a number of universities in Brunei, China, Indonesia, Malaysia, Papua New Guinea, the Philippines, Thailand, and Vietnam. Offshore courses are run in Brunei, Hong Kong, India, and Malaysia (http://www.dest. gov.au/tenfields/architecture/2ntu.htm, accessed on June 16, 2010).

CDU is also the largest research and development performer in the Territory and is now responsible for almost 25 percent of all research and development performed in the NT.

CDU is an example of a small university achieving excellent outcomes by adopting a structured and highly focused approach to research and research training. The University addresses issues that are of particular relevance to the Northern Territory including social, cultural, environmental and technological problems affecting development. Its research has a strong regional focus covering northern and central Australia and the western Pacific. The University has a distinctive focus on early career and new researchers, including those in non-traditional fields such as the vocational education and training sector. CDU has clearly defined its areas of research strength and research priority, which are reviewed on an annual basis. These areas include tropical environmental science, tropical plant science, environmental remote sensing, tropical aquaculture, tropical health and international business. Research in these areas is underpinned *by centres* which provide a focus for research among academics throughout the University and beyond. Its participation in *Cooperative Research Centres and an ARC Key Centre for Tropical Wildlife Management* also provides further opportunity for collaborative research and the development of postgraduate studies programmes. (DETYA, 2000, Innovative practices: Research and research training, p. 150, Higher Education Report for the 2000–2002 Triennium)

CDU and its precursor institutes have been engaged in extension service and education for rural development in Australia for a long time, and it has a close affinity with its local community. CDU has developed niches in areas of greatest importance to its local community and its development. These include tropical and desert region studies, international (particularly South-east Asian) law and business, and Indigenous research and education. However, as the only university in the NT, CDU offers a broad range of courses in traditional areas of study as well (http://www.ntu.edu.au/aboutntu/introduction.html, accessed on May 28, 2010).

CDU's Mission

As the only university in NT, and located in Darwin, the top end of Australia, CDU has conducted social, educational, scientific, economical, and cultural promotions of Australia's NT by its various activities and has set up its mission below.

The University will provide education, training, research, and related services locally, nationally, and internationally to support and advance the social, cultural, intellectual, and economic development of Australia's NT (http://www.ntu.edu.au/aboutntu/ourmission.html, accessed on May 28, 2010).

The Dual-Sector University Model

In recent years, cross-sectoral education and training in dual-sector universities in Australia have brought about effective outcomes and practices.

The CDU is one of the few Australian universities which offer a full suite of programs incorporating: TAFE and Higher Education (http:// www.ntu.edu.au/faculties/site/welcome2.htm, accessed on May 28, 2010). These universities characterize the amalgamated approach whereby the TAFE and higher education sectors co-exist within one institution but possess different internal structures. This has promoted educational benefits to create more opportunities for the articulation of units, credit transfer, the dissemination of new knowledge, and the multiple pathways for study. Clearly, with collaboration of this nature, both sectors have experienced economy of scale from the ability to share resources and facilities efficiently.

Research in the university is one way of supplying TAFE programs with new knowledge as well as applying knowledge in new ways. By working collaboratively, technical skills can be improved in horticulture, agriculture, numeracy, literacy, and so on in rural areas. Appropriate and efficient production systems can be developed, technologies developed and integrated, and agricultural productivity sustained and enhanced over the long term. For example, completed research projects at the Tropical Savannas CRC (Cooperative Research Center) at the CDU have enabled the results to be used for training packages in TAFE or in many instances, improved training programs (http://savanna.ntu.edu. au/education/index.html, accessed on May 29, 2010). Further, research results from the Center for Tropical Wetlands Management have led to the development of a suite of training courses ranging from management planning for wetlands to hands-on weed control. Some examples of projects include (http://www.ntu.edu.au/ctwm/annual.html, accessed on May 29, 2010):

1. Collaboration for better Aboriginal land management.
2. Overview of weeds in Aboriginal land.
3. North Kimberley traditional owners' land and sea management planning project.

4. Guidelines decision tools and education programs for sustainable grazing management of savanna woodlands in the Burdekin River catchment.

5. Grazing Land Management which will be available in various forms.

6. Biodiversity in pastoral lands to be available for delivery in workshops TS-CRC (Cooperative Research Center for Tropical Savannas Management).

CDU's Involvement in Rural Development

CDU, like other Australian universities involved in an extension and VET for the delivery of rural and related education and training, does not have a formal mission for extension services or to provide formal extension courses. In the NT, the Department of Business, Industries and Resource Development (DBIRD) have a function for extension services. Similar Departments of Agriculture in other states have a similar function. In the past five years, CDU in cooperation with the NT Governments has provided workshops and short courses on various topics to rural communities (In 15, 2003). Some programs run by CDU have been recognized as a good example of how TAFE and industry can work together with training packages to make more flexible and practical instruction for the students. In some horticulture training packages, for example, the horticulture knowledge (theory component) is delivered by CDU lecturers and experts from local industry, and the on-job component of the course is completed within normal work schedules with assessors visiting students in the workplace (Intuition, December 2001, p. 8).

There is a key center in CDU, the key center for Tropical Wildlife Management, which has a long-term involvement in wildlife management in northern Australia and the wider region through its research, training, and education in rural and remote areas. It focuses on human capacity building and on commercial benefits. Its major activities are to recognize and meet the needs of Indigenous people in wildlife management; support the efforts of Aboriginal people in maintaining culture through the enhancement of customary and contemporary wildlife use and management; enhance sustainability of commercial use of wildlife in biological, economic, and social terms; and promote better public understanding of the significance of wildlife use as an alternative to other forms of land use. The Center's training programs include both formal and informal education for undergraduates, postgraduates, agencies, and Indigenous communities. Another major focus is on working with Aboriginal people to develop commercially viable, wildlife-based industries. A research in conjunction with several Aboriginal organizations across Top end and Cape York has been conducted to examine the feasibility of local, small-scale commercial plant harvests (Intuition, Vol. 14, No. 2, April 2002, p. 12).

Another example for Indigenous community development shows that some courses CDU has delivered are short term and focus on practical contents and apprenticeship training in multidisciplinary studies. Other courses CDU provides for Indigenous people have a strong theoretical content with practical and field-based experiences. Students therefore have the opportunity to apply their theoretical knowledge with practice. "The course structure allows all students to integrate their work and research practice with theoretical insights the course offers" (Intuition, Vol. 14, No. 4, June 2002, p. 7). Other NT organizations conduct research involving Indigenous people; for instance, Danila Dilba Health Service, Menzies School of Health Research and the Aboriginal and Torres Strait Islander Commission, and so on.

CDU is investigating the use of satellite links to its regional centers, for example, regional centers at Katherine, Nhulunbuy, and Jabiru, which may provide better service (e.g., greater bandwidth) at less cost (Intuition, Vol. 14, No. 3, May 2002, p. 2). The connection to CDU is via NT Government's wide area network (WAN), other centers' connection will also occur soon.

Training Materials and Learning Methods for Rural Communities

CDU has delivered its training materials for rural communities, both internally and externally through internet, hardcopy materials, CD-ROM, etc. For example, CDU has cooperated with Territory Insurance Office (TIO) to run the University's Remote Area Driving Program to develop a set of programs, for example, a road safety video, driver training, licensing needs for remote and Indigenous communities for the purpose of making driving training more accessible in communities and helping to reduce road deaths among Indigenous people throughout the Territory (On Campus, Vol. 4, No. 10, June 2002, p. 2). The video is a set of segments, which covered cyclists, traveling in open road space, road conditions, pedestrians, alcohol and other drugs, obtaining a license, first aid and management of a suspected spinal injury, trip planning, and community road safety. The video is delivered with an accompanying workbook including an outline for small group training. There are also questions for discussion, suggestions for activities and role plays, and list of focus points. These CDU remote areas' training programs were established in 2000 and focused on driver-in-training and licensing needs in remote communities. Before that, there was no driver training for Aboriginal people based in remote communities. People in these communities drove cars without training and without a driver's license. Aboriginal people represent 30 percent of the NT population but account for 50 percent of annual road deaths (Intuition, Vol. 14, No. 5, July 2002, p. 9).

Some short-term courses or rural participating programs have been delivered by using face-to-face, tele-conference, and video-conference communication methods (Intuition, Vol. 13, No. 11, December 2001, p. 12). With online teaching and learning becoming the major part of university's activity to deliver knowledge and information for rural and remote areas, CDU has encouraged its staff to adapt online teaching modes. These three modes were defined by the Australia Commonwealth's Department of Education, Training and Youth Affairs (DETYA). They include: a web-supplemented program (mode A), in which participation online is optional for the student; a web dependent program (mode B), in which participation online is compulsory; and also a full online (mode C), in which there is no face-to-face teaching and learning in the program, all interactions with staff and students and other activities are integrated and delivered online (Intuition, Vol. 14, No. 2, April 2002, p. 10), which makes greater use of advanced remote area telecommunications for open learning, working in partnership with communities.

Education and training are long-term interests and set the future of any society, and CDU has proposed to make more contribution in rural areas by suggesting: "...Wider use of recognition of prior learning and the flexible delivery of programs to remote and regional areas; and the establishment of a Subtropical/ tropical agriculture school, based jointly at NTU (CDU) and James Cook University" (Queensland) (http://www.jcu.edu.au/cairnsinstitute/public/groups/everyone/documents/working_paper/jcu_128813.pdf, accessed on December 5, 2014).

Case Studies

Introduction

In order to collect first hand data for the case study, and learn the experiences in an Australian's university, I have carried out surveying, delivered questionnaires, and interviewed staff in appropriate centers and faculties. I also paid a visit to NTRC on December 8, 2001 and talked with the staff in this college, as well as visited and discussed with and interviewed staff in the target research centers for the study. The common activities I found are explained in the following text.

Northern Territory Rural College (NTRC) (Katherine Campus)

Location

The NTRC, which merged with the CDU in January 2000, is located at Katherine, 300 km south of Darwin, which is the heart of the beef cattle land, and a rapidly developing mining, agricultural, horticultural industry, and tourist region (Tan et al., 2001, p. 8). "The site of present-day Katherine was the junction point of the traditional country of four Aboriginal tribes—The Djuuan, Wadaman, Sjamindjund, and Dagaman people" (*1981 Student Handbook*, for One-year Certificate Course in Rural Studies, Katherine Rural Education Centre, Northern Territory Australia, 1981, p. 3). As one of the four campuses of CDU, the NTRC has excellent teaching facilities with well-equipped workshops, cattle handling facilities, stables, modern air-conditioned classrooms, and a library/computing facility located on its 4,000-ha main campus. It also has its own 700 km^2 cattle station situated 90 km south of Katherine at Mataranka. Full residential facilities are provided for those attending the NTRC and the CDU Regional Center in Katherine. The NTRC also provides a daily return bus service for students commuting between Katherine and the center (http://www.ntu.edu.au/faculties/site/schools/spi/campusntrc.htm, accessed on June 3, 2010).

Programs

NTRC provides training programs according to the needs of the individuals, industries, and communities of northern Australia. Its key functional responsibilities include:

1. Developing, in conjunction with industry and community bodies, an annual training profile which meets identified training needs.
2. Delivering quality training programs for people seeking careers in agriculture, horticulture, and associated industries.
3. Delivering appropriate training to clients located in areas remote from the NTRC, with emphasis on Aboriginal training.
4. Maintaining facilities for the delivery of training and residential services.
5. Providing a Northern Territory Certificate of Education (NTCE) Program which meets the requirements of the NT Board of Studies (Tan et al., 2001, p. 8).

Programs range from on-the-job and off-the-job training for disadvantaged youths, certificate level courses in agriculture, horticulture, to diploma and advanced diploma level courses relevant to the overall management of tropical beef cattle properties, horticultural enterprises, and the aquaculture industry (Tan et al., 2001, p. 8).

Short courses are also available throughout the year. There are designed to meet the needs of the community. Some examples are car care, chainsaw operations, fencing, four wheel driving, horse riding, horse shoeing, motor bike operations, record keeping, vehicle maintenance, welding and communication in the workplace. Sometimes, NTRC delivers an individual course; sometimes a few subjects have been selected from national training package depending on local requirements. For example, in August 2002, a group of students completed a 15 weeks Stock and Station Skills Course and began their new career in the rural industry. The course is for practical content with students completing horsemanship, cattle handling, health and safety, fencing, water supplies, safe use of chemicals and chainsaws and welding and maintenance training. Students who gained competence in all of these subjects will graduate with a certificate II in Agriculture (General) (On Campus, Page 5, Vol. 4, No. 15, August 2002, p. 5).

Agriculture in Katherine Area:

> Katherine is the commercial centre for some 80 pastoral properties in the Elsey, Gulf, and Victoria River districts and 40 smaller farms along the Katherine and Edith River. The traditional industry is cattle raising with associated horse breeding, but there is an emerging cropping industry and some quite successful flower, fruit, and vegetable production. (Student Handbook for Certificate Course in Rural Studies, Katherine Rural Education Centre, Northern Territory Australia, 1985, p. 18)

It has also long been recognized as an important horticultural center in the NT and some of NT's most experienced growers produce fruit and vegetables for local, interstate, and international markets.

Technical and Vocational Education/Training (TVET) subjects are integrated into the secondary school curriculum. The program's objective is to ensure that students wishing to complete Year 12 have access to a practical alternative to obtain the NTCE. Courses are designed to equip students for employment in the automotive, metal, and rural industries, as well as providing credits towards a variety of TVET courses.

Teaching Areas Provided by NTRC

The NTRC provides the following teaching areas:

1. Horticulture
2. Agriculture
3. Aquaculture
4. New apprenticeships/traineeships
5. Pest management
6. Farm chemical safety and application
7. Parks and wildlife management

New Opportunities

The incorporation of the Tropical Savannas CRC into the Katherine Regional Center and the NTRC have opened up avenues of opportunities for both training and funding for CDU, its students, industry, and the community at large. CDU ran its first full-time horticulture course outside Darwin in 2001. The small established mango orchard on the NTRC's 4,000-ha property at Katherine is used both as a training center and a commercial venture: Students are trained in pruning, picking, pest management, and crop care and at the same time the fruit is exported. Potential areas of expansion include improving the flexibility of the National Apprenticeships to deliver on- and off-the-job training and assessment, with emphasis on networking with industry to assist developing partnerships (Tan et al., 2001). NTRC also provided training before the job. This included horsemanship, health and safety, wildfire operation, advanced first aid, mechanics and welding, and so on (Intuition, page 11, Vol. 14, No. 2, April 2002, p. 11). In addition, potential exists for the development of strong research and global links. CDU in the Katherine region will aim to expand its course delivery in the areas of business development to include business, farm and pastoral management to meet the diverse needs of our rural community and regional industries (FI8, 2001).

As part of my field work for this research, I visited NTRC on December 8, 2001, and talked with the staff in this college, using a structured interview schedule (see Appendix 3).

Cooperative Research Center for Tropical Savannas Management (Tropical Savannas CRC or TS-CRC)

Australia's tropical savannas are shown in Figure 6.2 (http://savanna. ntu.edu.au/centre/, accessed on June 3, 2010) and cover almost a quarter of the continent. The Tropical Savannas CRC (Cooperative Research Center) was established in 1995. Its aim is to help ensure that this vast area is healthy and managed to provide long-term benefits (economic, aesthetic, social, and cultural) to those who use them and to sustain the biodiversity and habitat endemic to them (http://savanna.ntu.edu. au/centre/, accessed on June 3, 2010). The most important mission for the Tropical Savannas CRC is to make land management research more useful to various land managers and agencies across the tropical savannas. The work has been undertaken by the center through its research, extension, education, and training.

Research

The Center's research activities focus on Landscape Ecology and Health, Industry and Community Natural Resource Management, Regional Planning and Management, and Human Capability Development in

Figure 6.2
Australia's Tropical Savannas

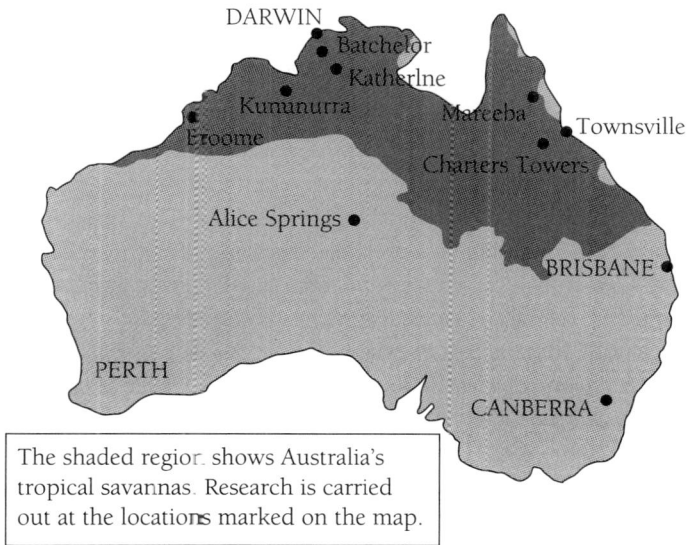

DARWIN
Batchelor
Katherlne
Kununurra
Broome
Mareeba
Townsville
Charters Towers
Alice Springs
BRISBANE
PERTH
CANBERRA

The shaded region shows Australia's
tropical savannas. Research is carried
out at the locations marked on the map.

Source: http://savanna.ntu.edu.au/centre/index.html, (accessed on June 3, 2010).

order to develop and intensify the use of natural resources in northern Australia. One of the Center's concerns on its research project is that the tropical savannas still retain the natural and cultural values, unlike the loss typical of southern Australia. On the other hand, the low population of the tropical savannas makes it difficult to develop a "critical mass" of researchers in many disciplines. The Center helps to overcome this problem by creating projects that bring together researchers from its 16 partner agencies spread across Queensland, Western Australia and the NT (http://savanna.ntu.edu.au/research/, accessed on June 3, 2010).

Education and Training

The Center provides educational materials at four levels, and delivery of courses, catering to a wide range of users. The activities include three parts: (a) Graduate certificate, diploma, and masters degree in tropical environmental management, which provides students with the opportunity to study, develop, and gain skills in the sustainable management of tropical ecosystems. All core units are delivered flexibly, to suit students' individual learning needs. Most of the material is presented in self-paced learning mode, using online, print-based and CD-ROM formats; (b) Student research projects require the work of PhD, honors, and masters students supported by the TS-CRC to contribute directly to the Center's research themes and projects. It also helps link graduate students directly with practice, helping ensure a practical research focus into sustainable land management research in the savannas; (c) Extension and VET, the key objective is to develop and deliver appropriate and relevant learning materials/packages to meet the different needs of the six sectors it serves: pastoral, aboriginal, conservation, tourism, mining, and military (http://savanna.ntu.edu.au/research/projects/extension_training.html, accessed on June 12, 2010).

The key center was funded by the Commonwealth Government, not the Territory Government. Its primary purpose is not to promote community development services, it is research, but one of the purposes can be the community development. For instance, most research programs are related to rural, remote and Indigenous areas and can contribute to community development.

Cases for extension, vocational education, and training

Learning packages, materials, video, books and case studies have been developed and are available, such as fire management, weeds, grazing sustainability and conserving biodiversity. A new degree in tropical agriculture has been setup to focus on three themes: Extension and Communication, Grazing Land and Animal Management and Healthy Savannas. For example, a video on weed management resulting from the research in Aboriginal communities was completed and distributed by 2001. The video aims to raise awareness about the spread of invasive weeds on Aboriginal land in northern Australia, with a focus on mimosa. A pocket guide on weed identification and control for two Aboriginal communities in the NT was proposed. Fifteen weeds of significance in the tropical savannas of the NT were selected. This project was placed on hold pending a review of the Center's extension program (http://savanna.ntu.edu.au/education/, accessed on June 12, 2010). Fire management is another example of CDU carrying out extension and training activities to serve the rural areas. For instance, the fire-management learning package and materials fall under the headings, such as, fire management book and case studies of practical fire management. These include the book *Savanna Burning: Understanding and Using Fire in Northern Territory*, which features a number of fire-management case studies assembled by this project. The property case studies were written to illustrate the book with real situations. Fire-management case studies have also appeared in the Grazing Land Management learning packages (http://savanna.ntu.edu.au/research/projects/extension_training.html, accessed on June 16, 2010).

These fire-management cases have covered the major research and management issues previously identified for northern Australia, and results of projects undertaken by the Tropical Savannas CRC fire program, which shows that: "the way fire might be managed to maintain diverse wildlife habitats and to maintain those patches that are rich in resources and crucial for maintaining regional wildlife population" (Intuition, Vol. 14, No. 6, August 2002).

CDU has also paid attention to the people from rural and isolated areas, and adopted strategies to improve access and participation for students from rural and isolated areas. The goal is to improve access and participation of students from isolated and rural areas, which includes the following activities:

1. Continue to enhance the programs offered through the University's remote campuses;
2. Continue to support the Isolated Children Parents Association;
3. Offer all CDU programs externally and on-line; and
4. Provide enabling programs on site to isolated students (http://mindil.ntu.edu.au/ntu/apps/ntuinfo.nsf/WWWView/Policy_333, accessed on August 9, 2010).

Many courses focus on the rural, remote and Aboriginal areas, for example, a Graduate Diploma in Midwifery requires that:

> unique to this course has been the development of remote clinical placement requirements, where students are encouraged to work within an Aboriginal community to enhance the theoretical cross-cultural learning. With a strong Indigenous population within the NT, the opportunity to learn from both health care professionals and community members in remote settings is of great value, and will translate into experiential learning of unquestionable relevance. (http://eagle.ntu.edu.au/NTU/APPS/CourseRe.nsf/0/D63D5BCDD57 FDB8369256DE60005FBB5?OpenDocument, accessed on August 9, 2010)

Center for Teaching and Learning in Diverse Educational Contexts (CTLDEC)

CTLDEC is located in the Faculty of Education, Health, and Science. Its research is conducted nationally and internationally in all aspects of learning for sustainable socioeconomic well-being.

The research areas cover formal, nonformal and informal learning, which focused on learning communities: organizations, schools, workplaces, communities, rural and remote regions, Indigenous contexts, on-line, civic, public and cultural as well as enablers of lifelong learning: literacy, language, leadership and management, social capital, community development, policy, and resiliency (http://ctldec.ntu.edu.au/welcome. htm, accessed on June 6, 2010).

The Center collaborates with state and territory government departments and agencies, professional bodies, industry and other groups and individuals in the provision of research and consultancy services in addition to continuing professional development for parents, teachers, trainers, and community educators. It also acts as the "clearing house" to publish and disseminate the outcomes of its research and consultancy activities.

The Center's activities have included teaching in isolated and small communities; compulsory and post-compulsory education and training; educational delivery; formal and nonformal education in the Asia-Pacific region; the education of Indigenous peoples; education at a distance; health education; social capital and community development; policy determination and implementation in remote circumstances (http:// ctldec.ntu.edu.au/aims.htm, accessed on June 6, 2010).

CTLDEC members are involved in a range of research and consultancy projects with significance for NT education. These projects have tended to emphasize rurality and/or remoteness, accompanied with complexities in educational delivery or access. Following are some successful stories from this Center.

Literate Practices in Indigenous Communities

The Aboriginal population occupied 30 percent of the NT's population and represented 13.4 percent of total indigenous population in Australia (ABS No. 3102.0), therefore, indigenous education becomes more and more important to the NT. The Center has concentrated its research in this area. For example, Professor Ian Falk is working with Inge Kral, Jerry Schwab, and Dorothy Lucardie from the Central Australian Remote Health Development Services to explore the links between literacy, social capital and community capacity in relation to culturally appropriate ways of improving health and well-being. The two projects were designed to complement each other in that each examined one whole Indigenous community's literacy practices in order to establish implications for more effective education and training to enhance health and well-being outcomes. One community was located in the Center (Utopia) and two in the Top End (Wudicupildyer and East Arnhem Land) (http://ctldec.ntu.edu.au/newspage.htm, accessed on June 6, 2010).

Discontinuities in Literacy and Numeracy Practices Between Indigenous Community Schooling and Urban High Schools

This case explains that Jennifer Rennie and a team of researchers from CTLDEC, Linda Ford, Peter Wignell, and Ruth Wallace are investigating how the literacy and numeracy skills of Indigenous students are affected when making the transition from community schools to urban high schools. The 2-year project, funded by the Innovative Links Project through Department of Education, Science and Technology, involved community schools and communities in the NT. An important outcome of the project will be the opportunity for community members to become involved and trained as research assistants (http://ctldec.ntu.edu.au/newspage.htm, accessed on June 6, 2010).

Workplace Literacy

Another program carried out by a CTLDEC member was to engage in the context of workplace literacy by examining what happened at a particular place and time when a group of Indigenous workers were confronted with a set of literate practice which were new to them. The location is a mine site in the remote Kimberley region of West Australia. The specific context is the development of workplace literacy training materials. The project was small in the broad scheme of things but involved negotiated engagement in a larger context of culture and subcultures (In 17, 2003).

Future Directions for Secondary Education in the NT

Professor Ian Falk, Dr Neville Grady, and other members of the CTLDEC team have been involved in undertaking a comprehensive report on secondary education in the NT, commissioned by the NT Government. The report will consider all aspects of secondary education provision by Government schools in these settings, and recommend future directions.

Input will be sought from the nongovernment school education sectors. CTLDEC members will be involved at all levels, in the reference group, project management committee, steering committee, making submissions, researching, analyzing data and writing. The team will meet with a range of key stakeholder groups including parents, students, school teachers, principals, and tertiary education providers, as well as representatives of business, community, the Australian Education Union and the DEET. With the focus on identifying priorities and strategies to improve secondary education in the future, CTLDEC members will be able to contribute their considerable knowledge and experience of secondary education and the NT (http://ctldec.ntu.edu.au/newspage.htm, accessed on June 6, 2010). This project will build on an earlier study, also commissioned by the NT government to suggestion reorganization strategies for the educational system as a whole.

From 1999–2001, a project was funded by the Committee for University Teaching and Staff Development (CUTSD) at NT University, now CDU to Facilitate flexible online teaching and learning.

The project began in 1999 as a joint project involving lecturers in the Faculty of Education and staff of the Open Learning Branch at CDU. The key elements of the project include (a) Environmental Scan evaluating the learning environments of students in urban, rural, and remote north Australia, to ascertain appropriate modes of electronic delivery of educational activities; (b) Development of on-line templates including the development of pedagogic and technical frameworks for four nominated online teaching strategies; (c) Four units of study, utilizing the four teaching strategies, are currently being developed. After pilot delivery and evaluation they will be available as exemplars; and (d) Professional development (http://www.ntu.edu.au/education/oll/, accessed on August 9, 2010).

Center for Indigenous Natural and Cultural Resource Management (CINCRM)

The CINCRM was established in February 1997 at the CDU with its initial funding from the Australian Government's Indigenous Higher Education Centers program. The Center is the research sector of the CDU's Faculty of Indigenous Research and Education. It is composed of staff and students in the faculty with the contributions from some other institutions such as Aboriginal Land Councils and other Indigenous organizations external to the Center's core activities, e.g., supporting and facilitating the participation of Indigenous people in research through project partnerships and steering committees. This network of Indigenous partner organizations plays a key role in assisting research activities and participation by Indigenous people, and the dissemination of research outcomes to Indigenous communities (FI9, 2001).

The key issue for the Center is to help Aboriginal people find the knowledge in an unbounded repository to reconstruct themselves. Mr Isaac Brown, the first director of the CINCRM (at that time it was called the Center for Aboriginal and Islander Studies), and one of the first Indigenous people who worked in higher education said: "We lost our economic productivity, our creativity. It was taken away and we're only just beginning to cope with that. We need to be able to reproduce the productive society we used to be, to reconstruct the cultural frameworks we lost. We need that to move on" (Intuition, Vol. 13, No. 11, December 2001, p. 6).

> The main missions of this Centre are to support Indigenous students and scholars to undertake research in natural and cultural resource management, and issues of sustainable community development affecting Indigenous Australians, particularly in curriculum development for Indigenous resource management, partnerships with Indigenous communities to incorporate Indigenous knowledge systems as a respected and valued body of knowledge in the Australian Higher Education system, and to establish and further refine research methodologies appropriate to Indigenous communities, so as to empower and inform Indigenous people in Australia and internationally. (http //www.ntu.edu.au/cincrm/aboutus/index.html, accessed on June 16, 2010)

The major achievements of research and consultancy are in the environmental sciences and the role of Indigenous knowledge and practice in the management of natural resources. The Center's researchers have undertaken collaborative research involving Indigenous countrymen and partners in other organizations in a number of key areas, such as: development of an integrated management plan for the Arafura Wetlands, traditional fire regimes in tropical northern Australia, sustainable use of trepang, Indigenous participation in tourism, Indigenous turtle management, etc. The Center has also paid attention to some strategic research themes. These are identified as: Indigenous governance and capacity building, and Indigenous ecological knowledge.

Besides its research activities, the Center publishes a series of discussion papers and a series of occasional papers and reports, workshop and conference proceedings, and occasional monographs.

The Center in cooperation with other faculties of CDU has a close connection with local industries and companies for Indigenous community development through training, research, and projects. The participating areas include:

1. Horticulture, especially as it relates to the delivery of information about growing fresh fruits and vegetables, developing orchards of exotic and native fruits, growing bush tucker and bush medicines for health issues, and developing lawn and parks/garden areas for dust suppression on Aboriginal communities, and so on.

2. Conservation land management issues, as it relates to the delivery of a new training package in community and outstation areas, specifically on how "Western Science" could complement Indigenous knowledge in the care and management of country. It also is concerned with how to upgrade skills for women on communities and provide basic scientific knowledge for Indigenous youth in diversionary and employment oriented programs.

   For example, the environmental company EcOz, which provides specific bioregional expertise in the wet/dry tropics and the arid zone of Australia, has pledged $1,500 annually to the CDU Foundation over the next 5 years in support of an Aboriginal student in the final year of their biology or environment degree, which has started in 2003 (Intuition, Vol. 14, October 2002, p. 10).

Government Commitment to Support CDU Financially and Administratively for Its Rural Development Projects

As the only university in NT, CDU has strong support from various government agencies financially, administratively, and project-oriented supports. For example, in 2002, CDU with a partnership of the NT Department of Community Development, Sport, and Cultural Affairs developed a project, called the Territory Housing Project, using the university's online learning facility to provide two programs: the Diploma in Front Line Management and the Certificate IV in Public Housing Management. The training programs have been delivered to staff in the workplace, using online learning technologies, facilitated by CDU staff and supported by workplace mentors from among Territory Housing staff. This is a government support project and would make contributions to NT government Indigenous affairs. The aim is to have staff in Territory Housing receive a formal accreditation training so as to improve frontline staff skills and capacity to identify innovative and creative solutions in service, which will better be able to meet the diverse needs of Indigenous clients. "In a first for Australia, this will be on-the-job training utilizing an e-learning program called 'Blackboard'" (Intuition, Vol. 14, No. 3, May 2002, p. 7).

In July 2003, CDU signed a new partnership agreement with the NT Government on Internet-based education for remote communities and a virtual DNA facility. The then CDU Vice-Chancellor Professor Ken McKinnon said the partnership agreement signaled a new level of interaction between the University and its community. "As a University for the Territory, it is critical that our intellectual resources, in collaboration with those of the Government, are brought to bear on the issues of most importance to the Territory." This Agreement includes 25 schedules based around four themes, which are: Increasing resident professional capacity to address Territory opportunities: Meeting Government needs; Reorganizing the University to better meet Territory needs; Enabling Indigenous social and economic development. The main activities focused on remote communities with the specific needs of the Territory such as, community development, conservation biology, natural resource management, and tropical environmental science, health, and diagnostics, as well as Indigenous social and economic development (http://www.ntu.edu.au/newsroom/stories/2003/july/partnership/index. html, accessed on July 1, 2010).

Another example shows the strong Federal government commitment as follows:

> The Department of Education, Training, and Youth Affairs has long recognized the higher costs of delivery of higher education at the NTU (CDU) and has provided operating grant funding accordingly. The NTU (CDU) has, and continues to receive, operating grant funding at a rate that is 20% higher than the national average funding rate per WEFTSU (weighted equivalent full-time student unit). The 20% loading is provided to offset the NTU (CDU)' remote location, its scale of operations, and its position with respect to higher education provision in the Territory. (http://www. dotrs.gov.au/regional/northern_forum/formal_response/top_end/higher_ education.htm, accessed on September 2, 2010)

In the VET sector, the implementation of Industry Training Packages (comprising competency standards, assessment guidelines and qualification outcomes) enables training providers to more easily develop flexible training programs. These programs can be tailored to better meet the needs of clients, including distance delivery arrangements.

Under the National Training Framework, training organizations must meet national standards before being registered. These standards include the capability to deliver client services, the recognition of prior learning, and the design and adaptation of training products.

These national quality assurance arrangements assist in providing the foundation for the wider use of the recognition of prior learning and flexible delivery for training, which would be driven principally by state and territory policies.

In the higher education sector, the Commonwealth encourages higher education institutions, including Batchelor Institute and the CDU, to further develop systems which recognize prior learning. Universities are responsible for determining their own academic entry provisions and are best placed to decide who may be accepted. The CDU and Batchelor Institute have been supported by the Commonwealth in developing infrastructure to deliver programs to remote and regional areas. The Commonwealth encourages them to explore innovative means of delivering educational support to students, including through flexible delivery.

For instance, the CDU received almost $1.6 million (2001 prices) from the Rationalization and Restructuring (R&R) program jointly with Flinders University for the development of tele-teaching links between the two universities and other linked distant sites. The idea of this development is to facilitate better delivery of programs to remote and regional areas. In addition, the Government has announced total funding of $3.2 million to facilitate bandwidth access to the CDU and six other regional universities under the regional Universities Bandwidth Project. The increased bandwidth will improve communication links, enable more cost-effective flexible delivery and will allow these universities to have access to Internet and other data services at a level comparable to their metropolitan counterparts (http://www.dotrs.gov.au/regional/northern_forum/formal_response/top_end/higher_education.htm, accessed on September 2, 2010).

Other Institutions' Involvement with Rural, Remote, and Indigenous Development Programs

Within Australia the university is not the only research institution, but private and government instrumentalities also conduct research and trials, for example, the CSIRO, Forestry Department, Aboriginal Board, Trade Department, Wildlife Conservation, Educational Department, Commerce Department and Indigenous Development, Trade Development Zone also have diverse research functions.

Research and Development

Research and development are a priority and are seen by the NT as the key issues to enable the region to make better use of resources both human and physical and to take greater advantage of opportunities. Some programs undertaken by the Department of Agriculture, Fisheries and Forestry, Australia (AFFA), focused on "expanding Australia's rural research and development effort, improving its efficiency and effectiveness by investing in high priority areas, and enhancing industry's international competitiveness through more effective uptake of research results" (http://www.dotrs.gov.au/regional/northern_forum/outline/the_forum.htm, accessed on August 15, 2010). There are 12 rural-industry-based Commonwealth Research and Development Corporation (R&DC) in the NT operating within the Agriculture, Fisheries, and Forestry. For instance, there is a mango flowering project, in which the productivity of Kensington Pride Mango which is low in the NT because of excessive and untimely vegetative flushing and unreliable flowering and fruiting. Two treatments to regularize flowering in the field will be evaluated. The first derives from CSIRO research which has demonstrated that flowering and fruit production of mangos can be enhanced by cutting a cincture around the tree trunk and applying a plant growth retardant, morphactin, to the cincture. The second flowering treatment uses paclobutrazol as a soil drench. The treatments will be evaluated in a multi-location trial (http://www.dotrs.gov.au/regional/northern_forum/outline/the_forum. htm, accessed on August 15, 2010).

There is a Commonwealth Scientific and Industrial Research Organization (CSIRO) Tropical Ecosystems Research Center (TERC) in Darwin (http://www.nt.alp.org.au/policy/newdirections/ndhedpp.html# key, accessed on August 10, 2010) and some other Cooperative Research Centers located in CDU involved in rural research and development programs. These are the CRC for Desert Knowledge, CRC for Aboriginal Health, CRC for Tropical Savannas Management, CRC for Sustainable Tourism, CRC for Tropical Plant Protection and Key Center for Tropical Wildlife Management (http://www.cdu.edu.au/research/centres_aop_ aos_foci/research_centres.html, accessed on August 19, 2010).

In Alice Springs, the Desert Knowledge Cooperative Research Center (DK-CRC) is another example. Established in September 2003, DK-CRC has become a venture for inland Australia's social, economic, and cultural improvement (http://www.atns.net.au/biogs/A001545b.htm, accessed on September 6, 2010). The center, with a core office in Alice Springs, has set up a network of researchers at 28 locations throughout the desert areas of Australia and worked with many educational and research institutions including CDU to develop:

1. Sustainable livelihoods for desert people, based on new natural resource and service enterprise opportunities that are environmentally and socially appropriate.

2. More viable remote desert communities that support desert people by developing attractive and efficient services.

3. Thriving desert knowledge economies that build self-sufficiency and minimize public subsidy.

4. Increased social capital of desert people, their communities and service agencies. (http://www.desertknowledge.com.au/Home. accessed on September 6, 2010)

The Center's research emphasized: (a) natural resource management for better livelihoods; (b) technical services for improved community viability; (c) governance, management and leadership for sustainable futures; (d) integrated systems for desert livelihoods to provide knowledge and outcomes in order to secure the sustainable future of Australia's inland. The Center engaged its diverse clients, which are regional communities in desert Australia, and including small business, Indigenous interests, local communities and government, large corporations, and state government agents. (http://www.desertknowledge.com.au/CRC/vision. html, accessed on September 6, 2010)

Education and Development

In the NT, the Commonwealth (the DETYA) and the Territory Government have delivered new apprenticeships in three sites: Darwin, Katherine, and Alice Springs, which is the only State or Territory till date where both governments have joined together to provide their functions to build on the success of traineeships and apprenticeships. This is a very flexible training package and focused on "employers in nonmetropolitan areas under the Rural and Regional New Apprenticeship initiative." "Industry Training Packages underpin the development of appropriate and targeted industry training in rural and remote areas" (http://www. dotrs.gov.au/regional/northern_forum/formal_response/top_end/ higher_education_part2.htm, accessed on September 6, 2010). They combine practical work with structured training and lead to nationally recognized qualifications. In some cases, students can begin a new apprenticeship while still at school. These qualifications can be delivered on-the-job, off-the-job, or a combination of both. New apprenticeships are now available in over 500 occupations. The programs include the qualifications from Certificate II training to Certificate III / IV training in an occupation that is identified as having skill shortages. Opportunities for groups such as Aboriginal and Torres Strait Islanders and young people initiatives are the focus of these (http://www.dotrs.gov.au/regional/ northern_forum/formal_response/top_end/higher_education_part2. htm, accessed on September 6, 2010).

Training Packages are a key resource for the delivery of the structured training arrangements of new apprenticeships. They have been developed for industry by national industry training advisory bodies, other recognized bodies or enterprises to meet the training needs of specific industries or industry sectors. Each provides an integrated set of nationally endorsed competency standards, assessment guidelines, and qualifications, and offers local employers and industries the flexibility to choose a registered training organization and negotiate an individualized training program suited to their needs and the needs of the new apprentice (http://www.dotrs.gov.au/regional/northern_forum/ formal_response/top_end/higher_education_part2.htm, accessed on September 6, 2010).

In addition, the Australian AFFA delivers training programs as a key issue for increasing competitiveness, profitability, and sustainability of Australia's agricultural industries. There include Agriculture Advancing Australia (AAA) package, which help rural businesses face the challenges of the future by becoming more competitive, sustainable, and profitable; and FarmBis program (AAA—FarmBis Australia), which provides assistance to national projects to enhance the business management skills of Australian agricultural and rural industries (http://www.dotrs.gov.au/ regional/northern_forum/formal_response/top_end/higher_education_ part2.htm, accessed on September 6, 2010).

For all training packages, there is flexibility. Training can be delivered on-the-job, off-the-job, during regular work, by student work experience or work placement, or (usually) by a combination of these methods. The ability to deliver training on-the-job and during regular work is an important element to ensure that training can be provided in rural and remote locations (http://www.dotrs.gov.au/regional/northern_forum/ formal_response/top_end/higher_education_part2.htm, accessed on September 6, 2010).

Government Commitment

The following case shows a strong government commitment. The Departments of Industry, Science and Resources; the Departments of Communications, Information Technology and the Arts, and the Departments of Education, Training and Youth Affairs jointly initiated a rural development package, which is a major beneficiary of this $2.9 billion package which includes (http://www.dotrs.gov.au/regional/northern_forum/formal_response/top_end/higher_education_part2.htm, accessed on September 6, 2010).

The $21.7 million New Industries Development Program which is specifically targeted at agribusiness and technology in rural Australia. (The New Industries Development Program has been expanded from a $4.6 million three-year program, to a $21.7 million five-year program) (http://www.dotrs.gov.au/regional/northern_forum/formal_response/top_end/higher_education_part2.htm, accessed on September 6, 2010).

Through the New Industries Development Program, Australian agribusiness will gain the business skills and resources required to successfully commercialize their business products, technologies, and services, and thereby generate significant growth in regional Australia (http://www.dotrs.gov.au/regional/northern_forum/formal_response/top_end/higher_education_part2.htm, accessed on September 6, 2010).

$155 million over five years for Major National Research Facilities with the potential for locating some of these facilities in regional areas (http://www.dotrs.gov.au/regional/northern_forum/formal_response/top_end/higher_education_part2.htm, accessed on September 6, 2010).

Other Activities

It appears as if many of the activities that are carried out by universities in China, for example, in terms of approaches to delivery of knowledge from universities to rural communities, such as visiting experts, setting up experimental bases, establishing extension and training system, leadership building, networking, farmers associations organizing, etc. are shared with other institutions in Australia. This is because of the different social, political, economic, and cultural environments of the two countries.

For instance, in AUH's case analysis, farmers' associations are organized by AUH. These are successful stories for bringing farmers into a learning society and use of science and technology. Similar societies in Australia and the NT are organized by farmers themselves or through government initiatives, for example, Dairy Farmer Cooperation, Goat Breeding Society, Pasture Protection Boards, Water Resources Commission, etc. Furthermore, AUH helps develop local enterprises to absorb surplus labor force from farmers in Hebei, China. In the NT, the Gas industry, Rail line, Trade Development Zone, Governments and private entrepreneurs do similar things using the surplus labor force from farmers (In 22, 2004 and In 23, 2004).

The above analysis and examples show that because of the different governments (China and Australia), and political, cultural, historical, and economic system, even CDU does some of things AUH did in China, but other organizations, institutions, or government agencies carried them out in NT.

Conclusion

CDU is a regional university with almost all the functions that a university should have, for example, teaching, research, and related services to support the social, cultural, intellectual and economic development of Australia's NT. Unlike the universities in China, the United States, and some other countries, where extension is one of the important functions to transfer knowledge and to serve local communities, especially rural areas, CDU, like other Australian universities, is involved in delivering rural and related education and training. They do not have a formal mission for extension work or provide formal extension. However, CDU is one of a number of dual-sector universities with TAFE and higher education. In the last five years, CDU has provided workshops and short courses to rural communities (In 15, 2003). In terms of knowledge demonstration, or extension, or serving for rural communities, CDU currently does not do many activities in this area, especially in higher education. However, the TAFE sector has programs serving local communities including Aboriginal people. CDU's mission is to serve local communities, which would include adult people and Aboriginal people in the communities. The standard way to deliver its units is through its external flexible delivery products, at degree and postgraduate level. At TAFE level, CDU staff have gone out and offered many whole programs, for example, for the Aboriginal people, horticulture staff offered certificate I and certificate II, and Certificate in Horticulture for Indigenous Community (CHIC) courses. There are short courses, where staff goes to the community for a few weeks to meet with a group of students. They may teach four or five modules, for example, how to use agricultural chemicals to grow crops. This program might use farm machinery and equipment. Staff implements the program in a concentrated block by traveling to communities several times a year to offer a whole program. Other programs consist of very specific modules, for example, for the Aboriginal communities and cattle stations, people learn how to use machinery to cut down trees and fix up the fences. Sometimes CDU staff go out to offer, perhaps one day courses on some subjects, particularly in TAFE, and that is at a practical skills level, using machinery, using chemicals safely, and so on (In 16, 2003). The above activities have been carried out by the Faculty of EHS (Education, Health and Science); Faculty of Law, Business and Arts has run TAFE level business management units, for example, Crocodylus World located in the faculty, which is a display and business initiative. It is a good example of a cooperative venture between the Crocodylus World business and the faculty. Other faculties also get positively involved with activities serving the rural communities.

CDU has substantial infrastructure to support rural research, for example, it has acquired a cattle station for commercial purposes, investing, and establishing high education capacity, to do research with government departments, provide formal, nonformal, and informal education for rural people, such as diploma, degree, postgraduate, education, delivering short courses, short time training, and so on.

An important way I think CDU could participate more in rural communities is to establish a lifelong partnership with rural communities, so that individuals have an opportunity for lifelong learning. These options should be flexible including formal and informal training. Training courses in rural areas could be many one-day courses. CDU should have a "skills' bank" for each individual in which the student banks their skills and knowledge over a long period of time adding up to certification. CDU needs to establish a long-term learning community in rural areas on a formal basis (FI10, 2001).

CDU also is a university with much cultural diversity in students and staff from different cultural backgrounds; for example, there are 262 students from 47 countries, even though some students are born in Australia, their original language and culture is not Australia. In CDU's Casuarina Campus there is a "Chinese Garden." Within it, a Chinese landscape, architecture, and culture have been developed. A bronze statue of Mr Tao Xingzhi (1891–1946), a famous Chinese respected and renowned educator occupies a key position within the garden. It was presented by Anhui Provincial People's Government of China, where Mr Tao was born. Mr Tao, whose name means "doing then knowing," is a well-known educational philosopher in China. He is known as "the great educator of the people" or "the masses education." Tao's principles of education progressed from the creation of an experimental school in a rural area named as Xiaozhuang to a national education reform movement (http://www.ntu.edu.au/mpr/mediareleases/mediareleases2002/july2002/educator.html, accessed on June 16, 2010). This has clearly shown the educational and cultural diversity of CDU.

Close linkage with Asia. This unique location, culturally diverse environment, and long history in dealing with Asia allow CDU to grow as a famous institution in this region and to be involved in many different programs. For example, almost all faculties, schools, and research centers have research programs, students, and staff exchange for the Asia partners.

Another important practice of CDU is combining theory with practice to benefit students. The CDU practice firm mentioned earlier, Crocodylus World, generously sponsored by Crocodylus Park (a private business corporation) is a fine example of this. This program uses simulation: "Practice firms have simulated workplaces in which students learn about business by doing it in a safe environment, allowing skills and abilities to be tested and developed" (On Campus, Vol. 4, No. 12, July 4, 2002).

Clearly, CDU has done a great deal in attempting to transfer a knowledge base from the university to the rural and remote areas, especially for the Indigenous communities.

As mentioned previously, CDU as a higher education institution in NT, which cooperates with other agencies involved in training and it has delivered many programs for the Indigenous communities. It has made contributions to capacity building of Aboriginal people and their representative organizations in fundamental ways, for example, at regional level, concerning economic, social, and political development.

As indicated earlier because of the isolation, climate, soil, and other conditions in the NT, development has been restrained. There is little agriculture in this broad territory, only a few examples in horticulture, fishery, and stock feeding. Therefore, there is not a big demand for the agricultural extension in this area. But CDU has delivered agricultural extension activities to East Timor, which recently became independent from Indonesia, and it now seeks assistant from CDU. In April 2002, Mr Flaviano S. Soares, head of the Agricultural Faculty of East Timor National University visited CDU and accepted a national agenda. The number one priority in East Timor' agriculture is to feed the population. With this in mind, CDU has educated specialists, helped and rebuilt the food economy, as well as introduced more productive agricultural techniques for farmers. CDU has participated in this project in a number of ways; its graduates visited farmers and spent time in villages where traditional subsistence methods are still used. CDU staff and students also train farmers in more productive agricultural methods in order to help them to participate in a modern economy. CDU has also worked with the staff from the Faculty of Agriculture, East Timor National University to replace the Indonesian curriculum with a shorter, more intensive curriculum, geared to meeting local community and industry needs. The teaching methodologies have also been changed with lecturing hours reduced, students required to develop their skills in group discussion, report writing, and self-directed study in the library (Intuition, Vol. 14, No. 3, May 2002, p. 8).

From the above analysis, directions and initiatives, there are four important issues for CDU in its knowledge transformation activities to rural and remote areas.

First, the university itself. There is an urgent demand for the university to concentrate its views on rural, remote, and Aboriginal communities. The demand is in technological extension, training, and building local people's capacity, in order to make benefits for them and improve their living standard and well-being.

These kinds of programs could combine with the university's teaching, research, and other academic activities so as to obtain mutual benefits for the universities and local communities. Besides teaching, learning, and research, the university's staff and students should become more involved in extension programs, especially at the higher education level.

Second, rural community participation is another important aspect for knowledge transformation. The rural development programs carried out by the university or other agencies must be attractive to the people both psychologically and economically to make more motivation for the local people in the community.

Third, government commitment and actual involvement are very important. Apart from financial support, the government's commitment for CDU to undertake the rural service program is mainly built on establishing partnerships or cooperative activities. Government priorities, for example, were given to those programs which concentrated on areas in rural, remote, and Indigenous education.

Finally, the programs should be more suitable and flexible for local communities, and more sustainable to make sure that they could be still active after the program implementer has left the program site.

The conclusion is intended to show how CDU is changing its mission for their rural development programs, and how these changes might relate to our expectation today of where rural development is going, and where it might be in the future.

7

Juxtaposition

Introduction

The previous chapters have already established initial comparability of the two case studies and the universities (Australia and China). Each has been reviewed for their roles in rural and community development for each country. The main aim of this chapter is to look for similarities and differences between the two case studies as required by the comparative research methodology.

Australia and China are quite different countries in many aspects, but at least one thing is very similar, that is the people, communities, institutions, organizations, and governments in both countries understand that education is a way to empower people for the improvement of their life, to facilitate society for sustainable development, and to promote the progress to benefit human beings. Education is a vehicle to improve people's knowledge level and skills, to understand other people and society.

Similarities

The Similarities in Rurality in NT and Hebei Province

1. The rural people's educational level both in Hebei Province, China, and in NT, Australia, is lower than that in urban areas in terms of school attendance, dropout, and the literacy rate, as well as educational years.
2. Economic conditions in both countries' rural areas still need much improvement. One of the significant methods is to train rural people and empower them to change their situation.
3. In both systems, people living in rural areas tend to be less well-off in terms of access to and provision of basic infrastructure such as education, health, transport, communication, and employment opportunities than their urban counterparts

In rural Hebei, generally speaking, educational levels in rural areas are much less than those in urban areas. Low level of infrastructures in schools, shortage of qualified teachers, and lack of financial support create unfavorable conditions for rural education, and for sustainable development of education in rural Hebei (Hebei Provincial Educational Report, 2002, p. 12).

It is also apparent that in rural Hebei, the economic situation is still backward, with agriculture, in most cases, being the main source of the family income. Effective delivery of new applied technologies for the rural population assumes a significant potential to change the modes of economic activities and performance of the rural population, thus contributing to the promotion of quality of rural life and community development. However, to bring about such expected changes, particularly at the grassroots level in the rural context, has remained insufficiently documented both in practical and theoretical perspectives. The communities have obtained new expertise for creating more productive means for income generation while the education institution has also acquired additional roles, apart from its formal training programs, in preparing the villagers for technology-based production. It is also necessary to persistently raise funds through multiple channels into the rural areas.

Some people in this area are still in a condition of poverty since they lack individual capacities. In this context, education has been perceived of as a critical media to materialize science and technologies into desired productivities and social wealth. Technology extension and quality will definitely exert an impact on the performance of the labor force and the status of the community development (Zou, 1996). The living standards of rural farmers have been identified to correlate closely with their education, especially vocational status of the rural population. A national scale retrospective investigation among 67,000 rural households revealed that the average per capita yearly income level ranged from 442.84 Yuan (illiterate background), 542.96 Yuan (primary education), 616.3 Yuan (junior secondary education), 639.35 Yuan (senior secondary education), and 740.9 Yuan (secondary vocational education) (Ying, 1993).

The urgent educational needs for the rural population in Hebei Province lie in vocational and technical education. Up to now, Hebei has succeeded in the provision of basic learning, through its decades efforts in massive literacy campaign and universalization of nine-year compulsory education program. The literacy rate of 85 percent among youths and adults and 85 percent completion rate at primary level is a marked achievement. However, the accumulation of vocational skills remains highly inadequate. A recent report reveals a disturbing fact that among the rural active labor force, 60 percent have completed nine-year schooling but have obtained only short-term vocational skills training. The number of qualified technicians available for every 10,000 farmers only amounts to 0.16 (Hebei Provincial Educational Report, 2002, p. 19).

In the NT, NT's rural areas or rurality is considered as remoteness. This definition has been recognized by Griffith (1992): "This uniquely remote nature of Australia rural communities is a significant factor that historically, has shaped the nature of rural education and in itself constitutes a challenge to educational provision" (Wyn et al., 2002). These rural, remote, and regional areas have problems associated with economic decline in many fields, such as education, health, communication, transport, and employment opportunities (FI11, 2002). It is clear that rural economy and rural development require a broad range of education, training, occupation, and skills. Despite the fact that the NT government has policies on providing a common standard of education across all school, rural education provision still faces difficulties in terms of educational standards, financial support, school maintenance, qualified teachers, and educational quality.

According to the Year Book Australia 1990, the rural population of Australia approximated to 15 percent of the total population, or up to

30 percent if the definition of rural includes the inhabitants of large towns (for instance, centers up to 100,000 people) (DEET Report of the Review of Agricultural and related Education, 1991, p. 23).

The NT's rural communities are shaped by very different economic, regional, social, and geographical forces. Life in different areas and towns is very different. Rural NT includes mining towns, coastal towns, regional centers, and Indigenous communities, which have their own tradition, expectation of, and needs for education.

In Summary: There are indications that both systems face a low educational status in rural areas, which constrains rural population and rural development. The people in rural areas receive less education compared to urban dwellers. Economic conditions in rural areas are inadequate and not well developed. There exist sharp contradictions between educational demand and inadequate financing, which also constrains educational development in these areas.

The Similarities of Educational Approaches Used by CDU and AUH and the Common Practice, as well as the Models of Effective Practice

1. Both systems have compiled different levels of learning materials and provided various training programs to meet learning needs in rural areas with an aim to foster rural development.
2. Internet access, distance learning, and other media have been used to serve the rural household.
3. On-site training and demonstration have also proved to be the effective practice.

At CDU Internet service and distance learning is a main approach used by CDU to carry out its community service programs. In some cases, specific training materials have been compiled and used to meet different local situations and different learning needs.

There is also a broad range of courses delivered by CDU for higher education and TAFE to rural, remote, and Indigenous communities, both internal and external. Courses are very flexible, short-term, long-term, and available from certificate level one to PhD. Lecturers and program facilitators often go to rural communities for on-site teaching and training. All of these activities are intended to create a learning community in rural, remote, and Indigenous areas.

In order to enhance community service in rural, remote, and Indigenous areas, CDU has created some positions and programs related to rural, remote, and Indigenous communities, for instance, an executive think tank of Indigenous leaders has been formed to help develop Indigenous research. Education strategies have also been organized. The think tank members include one professor from Australian National University, one professor from the University of Melbourne, and two others from the Northern Land Council, and the Central Land Council. This consideration is to appoint Indigenous academic and community leaders to assist in developing strategies for a university-wide approach to Indigenous development.

The main reason for this initiative is to foster CDU engagement with Indigenous communities, which is a way to provide and encourage more pathways into the university. The initiative also supports Indigenous students and staff, as well as developing how to enhance the understanding of cross-cultural issues through staff induction program and course offerings.

This is a practical model to expand the university's knowledge base and capacity for indigenous development, and widen the university's commitment to engage more fully with the Indigenous communities, as well as set up an adequate understanding of relevant cultural issues (http://www.cdu.edu.au/newsroom/stories/2004/february/thinktank/index.html, accessed on March 21, 2010).

In another initiative, CDU has appointed a Chair (professionship) on Desert Knowledge, which is located at its Alice Springs Campus, to focus research and disseminate knowledge on the scientific and social development needs of inland desert Australia. This appointment is intended to ensure research into desert knowledge and to deliver the benefits of this into desert communities, since "The health, well-being and ecologically sustainable development of desert communities and arid lands are essential to the Territory's development" (http://www.cdu.edu.au/newsroom/stories/2004/february/chairdesertknowledge/index.html, accessed on March 21, 2010).

In AUH

In rural Hebei, learning resources problems exist such as insufficient teachers, textbooks or training materials, classroom, and other teaching or learning facilities. AUH initiated a number of approaches to enable rural people to have more learning opportunities such as: establishing a village library and editing learning materials for an extensive coverage of knowledge and skills. Internet, telephone, local broadcast, and television have been used as a media to disseminate agricultural techniques and skills. Furthermore, university professors and other staff often paid visits to rural communities to work with local farmers, to deliver training courses, and to demonstrate agricultural techniques and skills. The important principle adopted is to create an environment in which participants could have self direction on how to obtain a new skill and how to use it. How to facilitate farmers to practice this skill and improve their life was another important concern. In AUH's training process, the participants undertake learning and training programs under the guidance of the experienced facilitators (university professor or staff). These are from various fields with the necessary knowledge to provide specialist information to meet the different learning needs throughout the delivery process. For instance, for each training or learning program, first, AUH creates "a demand" though its broad investigation, information, and motivation methods, and then meet "the demand" by various training and learning methods, so as to give high priority in meeting the interests and needs for the different target participants (FI6, 2004).

Apart from above-mentioned activities, like CDU, AUH has also established positions, and institution like the Rural Development College, Mountainous Research Center to motivate the university's professors and staff to be involved in the rural development programs.

In summary

Both CDU and AUH have developed effective educational approaches to transform the knowledge, technologies, information, and skills, to a large extent, to meet different learning needs for rural households. Furthermore, this has provided practical solutions so that knowledge reached those target groups.

The Similarities of Knowledge, Technology, and Information Transferring from University to Rural Communities for Development (the Transforming Models)

1. The two universities, CDU and AUH, followed almost the same principles in their transforming knowledge, technology, and information for rural development service, which is the

comprehensive integration of teaching/training, research and consultancy (extension or production). Sometimes the universities put all three priorities into one project, sometimes; they used them one by one.

2. The overall objectives for the rural development programs carried out by both university are to empower local people in rural communities, to build human resources capacity, and to create innovative approaches for the effective implementing of projects.

In CDU

As the only university in the NT, CDU has identified education, training, and research as its priorities and positioned itself as a place for comprehensive response to the different needs of students, industry, and farmers both for vocational education and training sector and higher education opportunities. Apart from these normal activities, CDU has also paid great attention to the rural community's service for a long time through training, research, and extension work, and has a close affinity with local communities. Research activities carried out by CDU are closely linked with the NT's unique environment, such research activities are: tropical environmental science, tropical plant science, environmental remote sensing, tropical aquaculture, tropical health, and Indigenous-related research. Also, TAFE programs focus on new knowledge as well as applying knowledge in new ways. By working collaboratively, technical skills can be improved in horticulture, agriculture, numeracy, literacy, and so on in rural areas. Extension services concentrated on the development of appropriate and efficient production systems, new integrated technologies, and sustained horticultural productivity, with an aim of education for rural development (Chapter 6).

In AUH

AUH used to be agriculture-oriented university. Traditionally, it focused on its teaching, learning, research, and extension on crop and animal production. In the beginning of the 1980s, with rural reform and evolution policies of China, AUH has redirected its mission towards the broader aim of supporting and participating in rural development. This can be summarized as "integrated development of agriculture, science/technology and education." This means AUH must use all means available to disseminate relevant scientific knowledge and productive technologies into the rural communities. The involvement of AUH in rural development initially started in the early 1980s with group and individual service for farmers in Taihang Mountain areas in order increase their income. The project initiatives are to disseminate the productive agricultural technologies by all educational means among villages, to enhance the capacity of the target groups, and to improve people's living standard.

The educational approaches used by AUH for rural development activities have been changed from its initial stage until now.

There were a number of approaches adopted by AUH in the Taihang Mountain project. These were:

1. Experts made a comprehensive system analysis, a holistic approach in considering the existing problems and potentialities of the Taihang mountainous area, so as to assist the village to develop an implementation plan in close response to the local resources.

2. Setting up of experimental bases, which were aimed at disseminating the agricultural technologies, which serve as focal points for demonstration. These bases aim to disseminate the agricultural technology as well as acting as service stations (like demonstration stations) for both technology and farm inputs.

3. Establishing a system of extension and training for agricultural technology and over-all quality improvement of the science and technology workforce.

4. Mobilizing human resources to provide various consultancy and service to farmers.

5. Building up/providing and strengthening a powerful system of leadership (prefecture, county, township, village, and university levels) to make up a complete body of policy-makers, administrative support, and implementing units.

6. Providing various training programs.

7. Networking.

   i. By joining efforts with a local seed company, AUH set up a strong marketing network for Chinese Cabbage seed in Gaoyi County.

   ii. To provide information and training facilities, AUH contributed 6,000 books and reference materials and other teaching equipment to Yongnian County Professional School and to Anping County Beigu Farming School.

8. AUH organized farmers in various technical associations to bring into play the initiative of farmers to learn and use science and technology, such as, Mushroom Association in Tang County, Chicken Association in LaiYuan County, Red Fuji Apple Association in Shunping County, Watermelon Association, Peach Association, Vegetable Association, Maize Association, etc.

9. Developing new technologies to maintain in response the forthcoming technology-based agriculture Chapter 4.

Summary

The similarities of the two universities are concerned with adjusting their programs to new and nontraditional topics, new teaching and learning models, new partnership, so as to play an active and constructive role in rural development. The models used by the two universities emphasize the importance of interactive, mutual learning between formal and informal knowledge or technology systems and they stress linkages with local community members or farmers so that they actively participate in any university's innovative efforts.

The Similarities of Government Commitment to Enhance University's Rural Service Programs from the Study

Transforming the university's knowledge base into rural communities to support rural development cannot be realized without government commitment. Several questions must be addressed:

1. What is the government's commitment to encourage universities to participate in rural development programs? How can the universities initiate and be supported by government? What recourses of the universities can be committed to catalyzing this? The above questions are similar for both universities, CDU and AUH.

2. The lessons from CDU and AUH have shown that government commitment to enhance the university to transform its knowledge base into rural areas for rural development is a critical issue.

In CDU

Generally, there is a strong commitment of government to CDU to serve rural areas. This is more so in 2004 as a result of the repositioning of the university under the McKinnon (the interim Vice Chancellor appointed to change CDU directions) initiatives regarding policy. The government's commitment for CDU to serve the rural development has been recognized by its research and training programs as well as consultancy or extension activities. Priority was given to the research programs whose designative areas are in rural, remote or Indigenous areas, or target people from those areas.

There are several centers established in CDU, such as Tropical Savannas CRC (Cooperative Research Center), the Center for Tropical Wetlands Management, The CINCRM, the center for renewable energy research, and so on, that are specially focused on the NT's rural and remote areas.

The NT and National governments have also committed money to support the university to carry out programs in rural, remote, and Indigenous areas. Generally speaking delivering programs on rural communities are very expensive and difficult to do without concrete funding.

The university needs the commitment of funds for staff travels, accommodation costs, and so on. Some other fundings are needed for students to travel out of communities to take training, for example, to Darwin, Alice Springs, or Katherine. Each year the Commonwealth Government gives funds to the Territory government for the vocational education and training programs, and then the Territory government distributes those funds to the educational providers, probably 85 percent of those funds come to CDU (In 21, 2004). For example, in health clinics, which are a part of national rural health training network, $170.6 million has been put by the Federal government for the rural medical training (In 19, 2004).

Another example of the government's commitment to the university (CDU) is that it funds a certificate program called natural and cultural resources management. Students can start from a certificate level one in the communities and, in theory, continue until they complete a PhD (In 19, 2004).

The rural development programs carried out by CDU sometimes need extra government support, for example, the Territory Government holds money for vocational and technical training programs. If CDU

carries out programs in this field, Territory government commitment, permission, and support are necessary. But higher education is
funded differently; the federal government funds higher education (In
19, 2004).

In higher education, the NT government provides less money than
the national government, but in TAFE more money has been provided
for external teaching and learning.

The NT government is very keen for the people to work in the
Aboriginal communities and is willing to support them financially (In
20, 2004).

In AUH

Similar programs exist in AUH. Also, government has a very strong commitment to encourage and support AUH to initiate its rural development programs. Policies, strategies, and methods have been adopted by
government and used to foster the AUH involvement in rural development program.

In order to enhance and improve the close linkage between AUH and
local communities, some provincial research centers have been located
in AHU to build cooperative partnerships among the higher education,
various government authorities, and rural communities with empowerment areas or enterprise community designations. This initiative focuses
on technical support, agricultural extension, economic opportunity, sustainable community development, community-based partnerships, and
a strategic vision for change (FI5, 2002).

This innovative program of government commitment provides
research funds to develop and maintain community, extension programs, and economic development projects relevant to their community.
The project creates strong information sharing and technical assistance
links between government authorities, rural development field staff, university, civic organizations and rural communities.

The roles and initiatives of the provincial centers focus on developing
capacity within selected educational areas to serve as resources centers
for community and technical and economic capacity building. These
provide continuous information, technical assistance and training for
empowerment areas, enterprise communities, other rural communities,
and the rural development field staff.

The centers have provided the following for the communities and university staff:

1. Technical assistance on community and economic development strategies and methods within communities served;

2. Training for local government, local technician assistants, empowerment areas and enterprise community leaders, as well as for community members and local farmers;

3. Assisting the university to sustain themselves as learning laboratories for comprehensive community development in rural areas with high levels of poverty;

4. Participating as integral parts of a national network of similar higher education institutions to share information, technical capacity, and experience (AUH, 1998, p. 172).

Some centers have been established to directly serve community educational needs. They reflect a long-term commitment of both the government and the community, and have many services already in place. They also, in many cases, provide the only affordable entrance to higher education for many rural residents.

These provincial centers located in AUH have shown strong commitment from governments to assist the university to carry out rural development programs.

Summary

Rural development is not the university's main mission; it is the government's responsibility for the human resources development or natural resources development. Thus government should commit resources to the university to ensure a desirable outcome in link with developed policy.

The Similarities of Community Participating in University Rural Service Programs

Community participation is another key issue to ensure that any rural development program can be successfully implemented and achieve expected objectives. CDU and AUH have paid a great deal of attention to have community participation during the rural development program implementation.

In CDU

The CDU's rural development programs have been recognized and been participated in by communities. However, the communities themselves want to be independent and prefer to have the people with knowledge and skills in their communities rather than someone come from outside. This is an activity that is motivated by the community. The community wants young people, with knowledge and skills in their communities without losing them. They fear that if they come to study in CDU or other universities, the community will lose them when they graduate (FI13, 2004; In 21, 2004).

The good and effective way to transform knowledge and skills from university to rural areas is for university staff (lecturers, professors) to go out into the communities, live with them, and find out what is really needed in the communities, as well as help them to build up those knowledge and skills (In 20, 2004). This educational approach also requires that communities have to participate in the program so as to achieve the expected objectives.

In terms of the motivation of community participation, when CDU delivered some training programs, for example, health programs in rural communities, those kinds of programs are very recognized by local communities. The reason why the local people and community members want to participate in their program is that the community is very interested getting young people educated and also the training program is closely related to the community getting more welfare (government payment), after program has been implemented (In 21, 2004).

In AUH

In the implementation of the Taihang Mountain development project, various considerations have been put into practice by the project team from AUH to ensure community participation. For example, for any project, the team identified some people who were interested in and enthusiastic for the project, including chairpersons, members of the administration, and key villagers. Those people who demonstrated strong commitment to the projects, and also the chairperson or administrative members ensure support for the projects. Other village members' participation occurs when their involvement is necessary at different stages of the program implementation. This involvement results in program ownership. Chapter 4 explained and illustrated this process, and indicated the critical roles that communities need to take part in together to accomplish educational and economical development in their villages (Chapter 4).

But in some cases it is not easy to get people together to participate in the programs, and also some programs have failed due to the lack of the community awareness, mobilization, and participation.

The analysis so far has been at a macro level to identify institutional capabilities. Similarities and differences at the university and governmental levels have been discussed and ground work for the establishment of models of implementation and suggestions for transformation regimes at various policy levels. However it is felt at this stage that a closer examination at the micro level or grass route level may highlight other issues not canvassed so far. Below is a series of case study material at the micro level to ascertain what may be necessary to action policy decisions and best practice in rural transformation.

Professor Yu Zongzhou: A Case Study

Mr Yu Zongzhou, a professor in AUH, has been engaged in rural development programs for a long time. He has successfully carried out many programs in rural and mountainous areas. But sometimes he met many difficulties. For instance, when he first time went into Qian Nanyu Village in Taihang Mountain areas, he prepared to help the farmers prune Chinese chestnuts, which is necessary for fruit trees to produce more fruits, but at that time, the local farmers refused to cooperate with him, and even were very rude to him. They said, "Come on, where the nuts would bear on, if the branches were cut down." Mr Yu Zongzhou facing the embarrassing situation, said: I talked to them and explained to them patiently, and finally I persuaded one farmer who agreed to have three of his apple trees to make an experiment for comparison. I only pruned one of them, and left the other two without being pruned. Before I did this, I promised the farmer if in the coming year the pruned tree would not bear much more apples than the other two, I would make up for the loss, and if more apples were produced on the pruned one, all of the fruits would belong to the farmer. The next year, the pruned tree had much more apples in very good quality, which are not only big but also attractive in appearance. This demonstration encouraged and motivated the farmers to learn knowledge and techniques.

Therefore, I opened up an evening school for farmers. I taught them how to cultivate fruit seedlings, how to prune fruit trees, how to manage them and how to deal with the soil, water and so on. Gradually, I transformed all sets of techniques on growing fruit trees to the farmers. Each of them has already mastered at least one or two practical skills. After that, they began to be richer and richer, at the same time they felt the importance of knowledge. So they began to attend various training classes and other development programs to further improve themselves (In 9, 2002).

Finally, almost all people in this village came forward to participate in the programs carried out by AUH, because they became aware of the common benefits to be gained in economic development from the projects.

Professor Sun: A Case Study

Another example happened in Su Jiatuan Village of Taihang Mountain areas, which was a well-known poor village. The farmers' literacy level is very high; almost all adults have finished primary education with basic literacy and numeracy skills. These skills can help them in their daily life and living. But it is not enough for them to improve their quality of life and raise their income; they need to develop their knowledge and skills. In 1987, they intended to grow apples. However, in fact, they had no knowledge and skills about how to grow apple trees well or how to manage the orchards. Consequently, the village leaders invited Dr Sun Jianshe, a professor in horticultural department of AUH, and designated him as the head of village to be in charge of the agricultural extension and management of fruit trees growing.

Professor Sun selected Wang Lanqing and another household as the demonstration model households. He looked after these two model households, and showed them the effective techniques step by step. In 1992, the fruit trees began to bear apples. The production in only 24 apple trees of Wang Lanqing's orchard sold at a good price of 9,800 Yuan. The income of another model household's apples amounted to 10,000 Yuan. Among all the orchards in the village, the apples in the 2 model households were the biggest and the quality was the best. This demonstration showed the importance of knowledge and skill in agricultural production, and it motivated farmers to participate in training classes for practical techniques, and to attend evening schools for cultural knowledge. These experiences initiated their income-generation opportunities, and farmers become more independent and flexible. After a few years, family's income increased. People of the village became aware of the advantages of new technologies. Active participation increased. In each family there was at least one person who mastered one or two technical skills. Under the guidance of Professor Sun, some local fruit tree experts have emerged, and the area of orchards has been enlarged. The output of fruits has been increasing in succession. In 1995, the total output of apples in this village has been up to 500 tons, which is worth 2 million Yuan. With only the apple growing, the income per capita amounted to 1,800 Yuan. Sun Jianshe had to be honored as a village head with a PhD title in the Taihang mountainous region. Owing to the apple growing with scientific techniques and management, Su Jiatuan Village has improved economically (FI5, 2002).

Summary

From the above descriptions, community participation is a critical issue to ensure rural development programs achieve the expected objectives. It is also necessary to make sure of the effective dissemination of knowledge, technologies, information, and skills to the grassroots from the university.

The important issue for the community members is that they should be interested in and have a motivation to participate. To ensure this becomes reality, university' staff carefully designed and planned the programs to make them more attractive to community members and more closely connected to local condition and to their daily lives so as to achieve the expected aims.

Differences

The Differences on Rurality in NT and Hebei Province

1. Hebei Province is an agriculture-oriented province, and the majority of the total population lives in rural areas. For most of people in rural areas, agriculture is the only income, which means that Hebei's rural economies almost fully rely on agriculture. Therefore, rural education has been considered as the key issue for national development. Whereas, the Northern Territory's rural areas have a small population, and most of the population live in urban areas, which means that the NT's rural economies are almost exclusively based on agriculture.

2. The NT's rural areas are very different to Hebei Province in economic, regional, social, cultural, and geographical aspects with unique characteristics, including mining towns, coast towns, regional centers and Indigenous communities. In rural Hebei, regional diversity is small, and there is no Indigenous population (only a few minorities in rural Hebei). Geographically, it is divided into mountainous areas and plain regions.

In Hebei, China

Hebei Province is an agricultural province with a huge rural population of 68 millions, 76 percent of them living in rural areas. Agriculture as the primary industry in rural areas serves most of the rural dwellers. Unlike their urban counterparts, in most of rural areas, agriculture is the only income for most of the population; therefore, the status of rural development has always been the primary concern of the Hebei Province. It is clear that, without rural development, without the improvement of rural people's living standard and wealth, without educational development in rural areas, the national or provincial development cannot be realized.

Administratively speaking, China has been managed by government respectively at the state, province, prefecture, county, and township levels. The village is a basic administrative organization of the rural community at the grassroots level (Chapter 3). Consequently, the villages become apparently the battle fields of educational intervention (Zhang, 2001). From this point of view, the focus for development must be on the rural society, on the development of rural education and on the growth of the rural economy as a whole, including its farm and non-farm components.

In the Northern Territory of Australia

The NT is the third largest territory/state but has the smallest population and low population density, which makes it very different from rural communities in Hebei, China. Darwin, a coastal city and the capital of the Territory concentrates the population. This metropolitan pattern has generated a variety of difference in rural and urban landscapes.

Traditional agriculture is in decline, and accounts for a far smaller proportion of the economic output, compared with mining and tourism. But unlike China, most of people in rural areas are working "in fairly universal, service occupations" (Wyn et al., 2002).

The Indigenous population in the NT occupies 28.5 percent of the total Territory's population, and represented 13.4 percent of total Indigenous population in Australia (ABS No. 3102.0). Therefore the Indigenous education and Indigenous areas development become more and more important in the NT.

Summary

The most apparent differences within both systems emerge in relation to each of the following: rural population, rural diversity, and reliance to agriculture.

The Differences in Educational Approaches Used by CDU and AUH

1. There are diversities between rural Hebei and rural NT, which presented as different learning needs, and different teaching and learning methodologies. Those diversities required the universities to consider and carry out relevant programs to meet the needs. For example, AUH is located in Hebei Province, which is an agriculture-oriented area in a developing country, therefore the most urgent needs for rural people, who make up the majority of Hebei's population, are up-dated information and technologies to assist them in having incoming generation so as to change their living standard. They want to know what to grow, what to raise, how to grow, how to raise, whom to sell to, and how to improve their economic productivity. All of these required the university to adjust its working emphasis in rural development programs.

2. AUH is a university in a developing country, while CDU is a developed country's university; therefore, funding for AUH is inadequate to maintain its long-term rural development programs. Alternatively, research funding from government and other agencies have been available in some specific areas.

3. Because of its cultural diversity and the Indigenous communities, CDU works across cultures to deal with students from different languages, cultures, and Indigenous background. The important approach in education and training is to be very patient with them, and to develop understanding of them, retain their culture, and build upon their previous learning. Thus teachers need to help, support, understand them, and not criticize them (In 19, 2004).

In CDU

CDU is a comprehensive university, although most of work on rural development is still concentrated in agriculture or agriculture-related faculties, rural people's needs and rural area development also concern other departments and faculties.

CDU has delivered short-term courses or rural participating programs by using face-to-face, tele-conference and video-conference communication methods (Intuition, December 2001, Vol. 13, No. 11, p. 12). With online teaching and learning becoming the major part of university's activity to deliver knowledge and information for rural and remote areas, CDU has encouraged its staff to adapt online teaching modes, all interactions with staff and students and other activities are integrated and delivered online (Intuition, April 2002, p. 10).

Consequently, a variety of ways of communicating its knowledge in terms of educational approaches is used but CDU as policy uses information technology or so called digital means, such as internet access, tele-centers and where ever possible. Compared with traditional media, this technology or information dissemination systems carry tremendous information with wide coverage, low cost, and timeliness. It also provides an all-round service with its large volume as well as its excellent verbal, sound, and image information.

CDU's rural community learning service is mostly focused on the individual, for example, when someone enrolled with CDU's external learning program, CDU will send the learning materials, and lectures will contact with them individually. Another characteristic is students are very scattered, for instance, probably one student in Alice Springs, one in Katherine, one in Japan, or one in Malaysia.

Learning activities or approaches consist of:

1. CDU's delivering knowledge to rural communities is very individually based. Both lecturers and students accept responsibility for the learning;
2. Creating a learning community on line; and
3. Lecturers going out of CDU to work with rural communities (In 20, 2004).

In AUH

AUH as a higher educational institution focused on teaching, learning, and research on crop and animal production. In the beginning of the

1980s, with a Chinese national rural reform policies, AUH has extended its mission toward the broader aim of supporting and participating in rural development with the groups and individuals service for farmers in Taihang Mountain areas for income generation and enhancing the capacity of the target groups.

AUH also provides one- or two-year diploma or certificate courses that prepare technicians for entry level extension work or entry level technician work with the public or private sector, in-service training programs for extension staff, farmers' training of short duration, adult training, and young farmer education and training (FI8, 2004).

Entrusted by Hebei Provincial Education Department, AUH implemented a program called "one village, one graduate from AUH," which means that each village is eligible to send a young farmer (of course, past the basic qualification assessment) to study in AUH free. After graduation, he or she has to return back to their hometown.

The key issue for AUH to deliver its rural development program is to establish a system of extension and training for the agricultural technology and overall the quality improvement of science and technology in the work force. In order to do so, some of the following approaches are used as:

1. Compilation of practical technologies based from actual research done on the site.
2. Technological training of farmers to adopt necessary technologies that require less work hands but with high production efficiency.
3. Organize a rational technical delivery service system (Consultative group) under the technology network.
4. Advocacy in setting up of various Professional Technical Societies (Fuji Apple Development Society).
5. Establishment of scientific and technology market of AUH (1988) linking university with farmers (vis-à-vis) for mutual benefit.
6. Selecting poor counties (10 out of 39 in the whole province) to set up experimental villages to carry out the poverty alleviation strategies.
7. Push forward some national projects and promote rural professional education, that is, Harvest Plan, Spark Plan, the Lia Yuan Plan, etc.
8. Training a large group of farmers to become chief members of an extension work force. For example, Beigu Farming School

with assistance from AUH experts trained 16,000 farmers among whom 240 farmers were appointed as farmer technicians by the county which made the farming output in 1993 raised by 56 percent higher than that of 1988 (Chapter 4).

Summary

Since the population in rural Hebei is much more than in rural NT, the educational approaches used in rural Hebei have concentrated on encouraging rural people in self-learning, helping each other, under the guidance of qualified staff, for example, various farmers' associations. Population's educational needs in rural Hebei, regional, economical, and agricultural diversity have also been factored into educational approaches during the delivering of rural development programs.

Cultural diversity and the Indigenous communities have been identified in rural NT. The cross-cultural problem requires the university to consider the target group difference in educational requirement and deliver the relevant programs.

The Differences in Knowledge, Technology, and Information Transferring from University to Rural Communities for Development

1. AUH is an agriculture-oriented university, in which its typical programs focused on natural science, agriculture, and agriculture-related social science, including farm management, agricultural economy, marketing, and so on, therefore its rural development programs.

2. CDU as the only university in the NT, its rural service programs focused on education/ training, health, Indigenous education, extension, and some agriculture or horticulture fields.

In CDU

Most of the programs carried out by CDU to serve rural communities are in adult training, external learning, distance learning in the areas of education, health, environment, horticulture, and other practical skills, as well as a variety of research and consultancy/extension (Chapter 6). For the research program, its primary purpose is not to promote community development service, but some programs can make a contribution to community development (In 21, 2004). For example, Tropical Wetland Management Center was funded by the Commonwealth Government;

its objectives are to promote wise use of tropical wetlands through cooperation in research, education, and training in relation to wetlands in
northern Australia. Therefore rural development is not this center's main
mission, but through its research activities, the contributions could benefit the rural, remote, and Indigenous areas.

In AUH

Hebei is an agriculture-oriented province; rural population is a majority
of the province, and agriculture is the most important or only means
in some area for most of them to live in. Therefore, most programs to
transform knowledge, information, and skills from AUH to rural society are through agricultural extension service, income generating, adult
training, and consultancy. As given in Chapter 5, it is apparent that the
fundamental way for the transformation from the practice of traditional
agriculture to the modern agriculture depends on the development of the
agricultural technology and the farmers' vocational and technical education. Hebei Province is a big agricultural province, with the rural population taking up nearly 80 percent of its total population. At the beginning
of the 1990s, the population with elementary education ranked 11th in
the country, with the illiteracy rate the 13th (according to the statistics of
the fourth census). Indeed, there existed a huge population not only in
normal illiteracy, but also in technological illiterates with only low-level
elementary education. They lack basic knowledge and skills on agricultural technology application.

With these urgent educational needs, AUH adopted the following
methods to transform knowledge, technology, and skills to alleviate
technological illiteracy and to promote economic development in the
Taihang Mountain area.

1. First, a large number of various training courses on technology
    and practical techniques closely related to different local con-
    dition and suitable to local economic development have been
    offered by AUH. Based on surveys and investigations of the natu-
    ral and social conditions in the area, and the following principle
    was established and strictly adhered to (Zhou, 1990, p. 58):

    > Priority should be taken over poverty alleviation while exploring and
    > making use of the resources in the mountains; getting rid of igno-
    > rance first before starting poverty alleviation programmes; education
    > programmes should be provided to get rid of ignorance; priority

should be given to those who have motivation when offering technical training courses. (INRULED, 2000, p. 15)

Under the guidance of this principle, AUH successfully transformed knowledge to mountainous areas and correctly dealt with the relationship of "transforming the mountain" and "eliminating ignorance" as well it raised the technological consciousness of the farmers. It is believed that as far as the regional economy is concerned, low-level elementary education is the basic reason for the poverty and backwardness in the mountainous areas. This is exacerbated by poor technological consciousness and technology illiteracy of the farmers. Therefore, many technical experts of AUH turned to be deputy county heads and township heads, who are responsible for local technical training and agricultural extension programs. Other professors and technical staff were invited to the grassroots level to popularize agro-techniques. As a result, farmers' attitude to scientific techniques was changed with the improvement of their scientific consciousness and the achievement of significant socioeconomic benefits. For example, farmers' annual per capita net income has increased from 70 Yuan in 1981 to 300 Yuan in 1985 and an economic benefit of 300 million Yuan was achieved on an investment of 7 million Yuan during first five-years of project implementation. Through this substantial practice, the farmers understand why "eliminating ignorance" should be ahead of "transforming the mountain" (Zhou, 1990, p. 56).

2. Second method, AUH carried out intensive and applied technical training to improve the farmers' technical skill. The comprehensive transformation and development of rural areas in the Chinese context required many technical personnel, but the rural areas lacked the technical personnel.

3. Third method, through its vocational school and vocational teaching center AUH trained qualified reserve personnel. It is found that by merely helping rural people out of poverty with science and technology as well as the applied skill that this only temporarily solves the problems. While improving the quality of life of the population with low-level elementary education, the farmers, can completely do away with such backwardness in the regional economy.

It is shown that every year a large quantity of elementary, secondary, and high school students join in the labor army of the farmers. These people, however, cannot adapt to the reality of the rural development due to lack of agricultural production knowledge and practical experiences. The vocational and technical center and agricultural vocational school can train people who qualified for the agricultural production.

For a long time, AUH has been helping the counties set up vocational and technical centers. It also helps the agricultural vocational school carry out teaching activities with the vocational technical institute as the center. For example, in 1996, it helped Shunping County establish the advanced agricultural vocational school, and according to the needs of Shunping County, selected specialities, curriculum; trained teachers for the agricultural vocational school; donated a great number of technological books, some teaching equipment and facilities for the agricultural vocational school.

4. Fourth method, professors or staff of AUH went off campus, work with farmers, and live in rural communities to demonstrate techniques and shared applied technical knowledge with them. This was universally and favorably received by the farmers and had significant achievements. Through this mobile guidance, demonstration and field operation, etc., the farmers have received the practical and effective techniques that brought in immediate benefit.

5. Fifth method, AUH initiated the development of the refresher institute and trained higher-level qualified technical personnel. To solve the difficulty of "recruiting and keeping the qualified people," with the support of the provincial government in Hebei Province, AUH in 1992 set up "the Refresher Institute of Construction in the Mountain and Old-aged Region in Hebei Province." It adhered to the principles "to assign the graduates back to their hometown, be lenient at the enrolment of the students and strict at the graduation of the graduates, stressing quality." Experienced teachers who had knowledge of the mountain development conditions were employed in the project.

For the recent several years, AUH has provided technical personnel from junior college who have become the fresh reformers for technical education in mountainous areas. Under their leadership, the farmers with low-level elementary education can receive the technical guidance at any time and be transformed from the technological illiterates to the technological "master-hand" in farms.

6. Sixth method, AUH disseminated the technology through its students' social practice activities during their vacation. AUH integrated the students' social practice with rural technological literacy and technical education of the farmers together. Every year AUH organizes over 40 students' groups and over 300 back-home groups in rural and mountainous areas. Through technology dissemination, technical consultation, short-term training and field guidance, etc., AUH is able to transfer technology and technical skill to farmers. AUH has improved farmers' technology literacy and consciousness and enabled them to apply or make use of technology in their farming practices. Those social practices, on the one hand, have given agricultural students actual exposure during their courses that inculcates love for the land and service for the farmers. On the other hand, it has facilitated the spread of technological conscicusness among farmers in rural and mountainous areas which improved their technological literacy (AUH, 1995, 1996, 1998; Wang, 1998).

Summary

From the above discussion, some differences can be summed up as in the following paragraphs.

China is a developing country; Hebei is an agriculture-oriented province, and AUH is an agricultural university; furthermore, the majority of the population live in rural areas, therefore, the urgent educational needs for rural Hebei are agricultural knowledge, technology, information, and skills. Concerning this unique situation, AUH has successfully transformed this kind of knowledge base into rural areas. It is evident that AUH has made its due contribution to rural communities, and farmers have reduced poverty and found a way to prosperity. AUH experience suggests that only when every rural laborer shakes off "technological illiteracy," will the influence be profound and total, because farm productivity cannot be actually raised until the farmers' technological consciousness and knowledge skills are upgraded and or duly improved.

Australia on the other hand is a developed country. The NT's rural population remains very small and is less dependent on agriculture.

However, the urgent educational need that is to be done for rural NT is the development of Indigenous communities.

The Differences of Government Commitment to Enhance University's Rural Service Programs from the Study

1. Given the view that rural development is not the university's main mission, it becomes the governments' responsibility for human resources and natural resources development. Therefore, governments have made the commitment, either to put some funds or to deliver some special policies or strategies in this regards.

2. China has a long time centralized, well established, and very organized government. In another words, almost all rural development activity needs the support, commitment, and financial aid of government. Australia is different. Its constitution of government is about 103 years old. This constitution of the Australian federal government in terms of education is quite different. Sometimes education is decentralized, and education becomes a state/territory government responsibility.

In CDU

The government's commitment for CDU to carry out rural service programs is, in most cases, to establish a partnership or with cooperative activities. For example, CDU, in partnership with the NT and Commonwealth governments, is developing an intensive implementation and research plan to expand an innovative literacy program in the Territory, including remote and Indigenous communities. This program is designed to accelerate the literacy skills of marginalized learners who have failed to make the appropriate literacy gains in school and/or who are in acute danger of falling behind the Territory average. The expected outcome is that the intervention helps students to gain literacy skills at a much higher level. This is also the first Australian state or territory to open up its schools to the research and development work required to mainstream the program, by the NT Government, working in concert with CDU (http://www.cdu.edu.au/newsroom/stories/2004/march/literacyprogram/index.html, accessed on March 21, 2010).

Both Commonwealth and Territory governments have committed CDU to deliver both higher education and vocational education. The commitment includes enhancing its role in meeting the social and learning needs of Territories, and to consider how this role as the leading educational provider in the Territory can be further developed (http://www.dotrs.gov.au/regional/northern_forum/formal_response/top_end/higher_education.htm, accessed on September 6, 2010).

In Australia, "government services have accepted major responsibility for extension services" (Donald, 1968, p. 11).

In AUH

The Taihang Mountain comprehensive development and transforming program was initiated in the late 1970s by AUH, along with the Chinese social and economic reform in rural areas. From that time on, the different levels of governments (central, provincial, and local) have made a long period of commitment and intervention. China is a centralized country with very strong government leadership. For example, in 1979, Hebei Provincial government made a decision to develop the Taihang mountainous region by means of science and technology. It started as a research project, which was named the "Comprehensive Research and Exploitation of Taihang Mountainous Area in Hebei Province." Committed by the governments, AUH undertook the responsibility for this project and organized professors, teachers, technicians, and hundreds of students to initiate the comprehensive exploitation (Zhou et al., 1990, p. 11). From that time on, the local government affirmed and set a high value on the exploiting of the model of this mountainous area experiment. Thus, in 1981, a key project called "Research on Exploiting Taihang Mountainous Area in Hebei Province" was established and was brought in line with the State Plan. Hebei Provincial Government organized some local institutions and cooperative units to accomplish this key project. Again, AUH bore the technological responsibility.

At the beginning of 1987, the Chinese Ministry of Education, in cooperation with the Hebei provincial government, decided to carry out a pilot program on comprehensive reform for rural education in Yangquan County, Shunping County, and Qing long County in Hebei Province. In 1988, based on the successful outcome of the pilot experiences, and the approval by the State Council, "the Prairie Fire Program," which means that "a single spark can result in a prairie fire," was put into practice and implemented all over the province. The key issues of the program are to integrate agriculture, science, technologies, and education; to carry out educational programs of the applied technique and management knowledge closely connected with the local communities; to train a large quantity of new-style rural constructor in order to speed up the agricultural development; and to establish a group of townships to help the farmers become well-off and promote agricultural development through education and technology transformation (INRULED, 2000, p. 31).

In 1990, National Instructive Guidelines for the Experimental Areas of Comprehensive Reform in Rural Education (1990–2000) were issued by the Ministry of Education to stipulate that "higher educational institutes, special technical schools and research institutes will be organized to participate in the education reform and economic development in the experimental counties. The provincial and local governments should designate the relevant higher educational institutes, special technical schools, and research institutes to connect with and support the work in the experimental counties," and "encouraging the cities to support the rural area," etc. (Lu, 1996).

The Chinese government committed higher educational institutes to serve rural development through the ways mentioned below.

To coordinate the counties and townships to formulate the social and economic development planning and programs; to help the village leaders and farmers change their ideas and concepts; to implement "the joint development project of production, learning and research"; and to help the rural area carry out various forms of education, train the qualified personnel, and improve the quality of the laborers (Lu, 1996).

On December 24, 1998, the Chinese Ministry of Education adopted the "Action Scheme for Invigorating Education Towards the 21st century, and emphasised that Higher Education Institutions should give full scope to their advantages in the state innovation systems, striving to promote innovation in knowledge and technology…; strengthen agriculture and work in rural areas … (Chinese Ministry of Education, 1998).

The Green Certificate program is another example of government commitment. The Green Certificate program was initiated by Chinese Ministry of Agriculture in 1990. In 1992, the State Council adopted a document of integration of agriculture, science and education, and improvement of rural economic development. In this document, gradually establishing the Green Certificate has been emphasized. Generally speaking, The Green Certificate is a national program for farmers conducted by the Chinese Ministry of Agriculture as well as an award given to farmers who have undergone technical training and who have proven potential and capability to extend their acquired practical skills. They must also have some management and supervision capabilities.

The farmers who obtain the Green Certificate have many benefits, for example, it is easy for them to have a loan from the bank (http://www. lzisti.net.cn/kjxn/lszs/lszs.htm, accessed on March 16, 2010).

Summary

AUH's rural development programs have been carried out under coordination, leadership, and commitment of government at all levels. Apart from funds from government, special policies or strategies have been also adopted by government. Therefore, without government participation, leadership, and coordination, it is very hard to carry out rural development programs successfully, especially in developing countries, like China. Mr Zhou Zhihua has shown that governments at all levels have played very strong leadership in any development programs in rural areas (Zhou et al., 1990, p. 48).

CDU as a higher educational institution has developed and participated in rural development programs. There are government (federal, territory, and local) commitments during the various project implementations. The federal government commitment is to ensure that CDU is a strong university to deal effectively with meeting the needs of NT. With the NT government formal partnership agreements between CDU and NT government or memoranda of understanding have been set up, and some activities have been carried out under these agreements. For example, the EHS, CDU has come down to the rural areas, carried out rural and regional participated research, consultancy and teaching or training work. This work not only helps rural communities and rural development, but initially it also identifies the future things that need to be done in rural development areas (In 18, 2004).

The Differences of Community Participating in the University Rural Service Programs

1. Community participation is a core element of a rural development program; the AUH rural development project is based on the principles of community participation, which assert that community participation is fundamentally required to achieve rural development and ensure the expected outcomes achieved at the local level.
2. The consciousness for participation in rural Hebei is less than that in NT.

In CDU

CDU's rural development programs have strong participation from people in local communities. For example, there is a research program in rural literacy development which is carried out by CDU which includes talking to rural community members including Aboriginal people and to identify what kind of English literacy they used. The research staff employed local Indigenous assistants. Their participation was as research participants in the project, they got paid, they were engaged to research and they got the training from research team how to do interviews, what to do after that and so on (In 18, 2004).

In AUH

Community participation requires going beyond consultation to enable community members to become an integral part of the decision-making and action process. It reflects the need for the development of more active communities in their own right: people seeing a need and acting upon it, for example, as advocates, pressure groups or self-help groups. Community participation draws on the energy and enthusiasm that exists within communities to define what that community wants to do and how it wants to operate. In rural Hebei, a few years ago, rural people were not willing to participate in the programs carried out by AHU because of the lack of the consciousness of technology, and also they had no confidence in their learning. For example, when AUH implemented the project of Revitalizing Villages through Science and Education, a professor had spent his time living with the farmers in Chaichang Village. When the professor conducted the first technical training class, no one came at the very beginning although the leaders of the village had announced the news several times. After further motivation by the leaders from door to door and under the condition promising to pay those who would attend the training, eight villagers came at last. Through their training, these eight villagers realized the importance of applying science and technology in their agricultural production, and they began to encourage others to attend the training class voluntarily. Gradually, more and more farmers came to classes, even those farmers from the nearby villages participated (In 5, 2002).

In summary

Since the most rural development work carried out by AHU is an agricultural extension work, the participating members are farmers. If they found that the project is closely linked with their income generation, and they will have benefit from the program, they are willing to participate. Their aims to participate are to find answer for many of the questions commonly asked by people when they get involved in agricultural activities and development initiatives.

Conclusion

The common concept apparently is that the universities have a key role to play in rural development through their teaching/training, research, consultancy/extension, or field work. University people must be aware of this and keep doing it effectively, efficiently, and creatively working to serve rural development. However, the university itself cannot play such a role if there is no policy support from governments. Failure to work with local people is another barrier to successful implementation. Networking with other institutions is also a critical requirement for rural development.

There are still some critical issues which are hard to juxtapose and hard to say that some kinds of work in one university are better than that in another since the two countries and two universities have a great diversity.

Main issues related to the university's rural development programs are as follows:

1. Rural diversity in rural NT and in rural Hebei;
2. The educational approaches used by two universities for rural development programs;
3. Models used to distribute the knowledge, technology, and information from universities to rural communities for rural development;
4. Government commitment to support and enhance university's rural development programs; and
5. Community participating in university's rural service programs.

There are three main hypotheses or questions:

1. The roles of university for rural development (why a university needs to serve rural communities);

2. Effective model for transforming knowledge, technology, information, and skills from university to rural communities (what activities should be used by universities to carry out rural development programs);

3. Innovative approaches undertaken by university for rural development (how are these models realized in practice).

If I examine the findings from the cases discussed under a IF–THEN regime, general conclusion can now be stated, if certain features hold as true then certain outcomes for rural transformation hold also.

1. The roles of university for rural development (why a university needs to serve rural communities)

If

   i. Universities, especially agricultural universities in developing countries have clearly identified that rural development is their main mission. Universities transform their knowledge base from research and apply it into rural areas.

   ii. Different levels of government have committed strongly to support universities to deliver rural development service both financially, institutionally, and with relevant strategies and policies.

   iii. Communities have paid great attention to the programs carried out by university for rural, remote, and Indigenous areas development.

Then

   i. University has a key role to play in rural development.

   ii. University cannot play such a role if there is no policy support or financial aid from governments.

   iii. A university's rural development program cannot achieve the expected outcomes if it fails to work with other institutions concerned to form a network serving rural development.

2. Effective model for transforming knowledge, technology, information, and skills from university to rural communities (what activities should be used by universities to carry out rural development programs).

If

    i.   Establishing demonstration communities.

   ii.   University's professors and staff are willing to go out of the campus and spend time and live in rural community.

  iii.   University has set up a network and build up a partnership with relevant institutions and organizations to share resources so as to transform knowledge, technologies and skills into rural communities.

  iv.   University's training, research and extension programs have been closely linked with the local needs and university has considered any specific conditions and situation in the program target areas.

   v.   Apart from the university's contribution for rural development, the university itself has also grown while it serves rural development.

Then

    i.   The university's rural development program can be more successful, effective, active, and efficient.

   ii.   Local community members are more interested in participating in the program.

  iii.   Expected outcomes can be reached.

3.   Innovative approaches undertaken by university for rural development (how are these models realized in practice).

If

    i.   Digital technology, Internet access, and other simple and effective media have been used by the university for its rural education and agricultural extension. Digital infrastructure has been extended from urban to rural areas. Then efficiency in terms of cost, staff's time, and learner's achievement will be much more increased.

   ii.   Organizing community members into various technical or learning societies, associations or other NGOs under the guidance of university staff. Then a learning society could be created.

iii. The university–community partnership has been established and using package contract approach and establishing joint ventures.

iv. Encouraging university student volunteers, especially agricultural university students, to launch social practice work and other practical courses in rural areas.

v. Setting up rural, regional, or night training and consulting centers.

vi. Training and encouraging a large group of community members to become backbone members of extension work force.

Then

i. New findings and new skills according to practice needs can be put into the communities for their development.

ii. Ensuring that all efforts are applicable, appropriate, and necessary for rural communities.

iii. Community members' ability and capacity can be empowered as well as a learning society can be created so that long term benefits can be achieved.

iv. The rural development projects can be more sustainable when the project implementers leave the project sites.

v. Apart from serving the rural development, university itself can also be developed; university staff and students can learn from farmers get benefits from implementing the programs.

8

The Common Factors
in the Cases

Introduction

In this chapter, some common factors are reviewed and compared. The idea is to have general understanding about the two systems and to trace how such understanding is reflected in the educational provisions of both study sites.

Comparison One: General Factors of Population, Rurality, Government, Economy, Language, Geography, and Education

Population

In 2000, the total population in Hebei, China, was 67.44 million, which ranks as China's fifth largest province. Of the population, 81.02 percent lived in rural areas, and the population density was 395.3 persons/km² (*Chinese National Statistic Handbook*, 2001). While, in the NT, the population in 2001 was about 200,000, the smallest population juris-diction in Australia. Less than one-fourth of the population lived in rural areas, and the population density was 0.1 person/km². Therefore, in Hebei, total population and population density are much higher than that in NT (Chapters 3 and 5).

In terms of rural people's dispersion, in Hebei, China, normally hun-dreds or thousands of households lived in a village, while in rural NT, people lived separately, and scattered throughout the vast landscape.

It appears that rural population and population density in Hebei, China, are much more than those in rural NT. Therefore, educational needs and provision of services as well as demand for rural development are far greater in rural Hebei.

Rurality

In rural Hebei, the economic situation is much lower than that in the urban areas. Most people rely on agriculture as the main source for fam-ily income. Educational provision is weak. School's attendance is low, dropout rates are high, and literacy rate is low. The basic infrastruc-tures of education, health, transport, communication, and employment opportunities need to be improved greatly. Some people in this rural area are in a condition of poverty since they lack development resources.

Education is one of the main measures to empower the people and it appears to be in urgent need. Effective delivery of new applied tech-nology and skills, promotion of quality of rural life, and community development are becoming more and more important in rural Hebei. All kinds of educational innovative intervention appear to be necessary (Chapter 7).

In rural NT, the urbanization rate is quite high. There is about three quarters of the total population, who live in urban areas (Chapter 5). There is almost no really "rural population" in the NT. Rurality is con-sidered as remoteness (Griffith, 1992). However, rural NT has suffered as a result of the lack of resources in education, health, communica-tion, transport, and employment opportunities. However, there is a wide diversity between the rural and urban areas.

China is a developing country; Hebei is an agricultural province, where most of the population live in rural areas and fully rely on agri-culture. Australia is a developed country, and diversity between rural and urban areas is small, but unlike other parts of Australia, the NT has faced many developing imbalance problems, especially between rural and urban areas.

Government and Administrative Structure

China has a centralized government, and its governance is based on a four-level structure and divided into provincial, prefecture, county, and township administrative units. In rural Hebei, the basic administrative unit is a village, including a natural village and an administrative village (may include a few natural villages). Generally, the size of a natural village is between 50 and 3,000 people. These villages are normally related to the geography of the area, the higher number in plains and lower one in mountainous area (Chapter 3).

In the NT, generally speaking, there are three levels of administrative units: federal government, Territory government, and local government. The local government includes city council, township council, and village council. On July 1, 1978 the NT became a self-governing territory. However, distance from urban centers is a problem for government (Chapter 5).

The key differences between China and Australia of government functions: China has a long time centralized, well-established, and very organized government. Many social aspects are government responsibility, whereas, Australia has had a democratic government for about 103 years. The NT has had self-government for about 26 years.

Economy

In 2000, the GDP in Hebei Province was about 507.63 billion Yuan, equivalent to 61.83 billion US$ and ranks sixth in China. GDP per capita is 7,527 Yuan, equal to 916.8 US$, and ranks 11th in China. These statistics show that the economic situation in Hebei is in the upper-middle level of economic development in China. Hebei has a long history of agriculture, and it is one of the main agricultural production areas in China (Chapter 3).

The economic development of Hebei accelerated after the "Cultural Revolution" and with the national "Open Door Policy" and reached an annual rate of 10.6 percent growth of the GDP.

In the NT, the economic capacity is small and only occupied about 1.3 percent of national GDP. In 2001, the Territory's GSP was valued about AU$ 7.45 billion. Unlike rural Hebei, there is very little farming to contribute to the economy. The economy relies on abundance of natural resources and mining, tourism, national defense, and a short distance to Asia.

Establishment of self-government in 1978 has given a positive opportunity for the economic development in the NT. In 2001, the GSP has increased 10 times compared to 1978 (Chapter 5)

It is very clear that both Hebei, China, and the NT, Australia, have experienced a fast economic development in the past 20 years, and strong potential for further growing and development. The scales are vastly different. Furthermore, agriculture has played an important role in rural Hebei, whereas, very little farming has existed in rural NT.

Geography, Language, and Cultural Diversity

In the center of the North China Plain, Hebei Province embraces two big cities, Beijing (the national capital) and Tianjin, with a total area of 187,700 km², which ranks 14th in land resources and about 0.2 percent of total area in China. The landscape consists of sea, mountain, plateau, plain, and wet land. With a temperate continental monsoon climate, Hebei Province has been divided into four seasons, spring, summer, autumn, and winter (Chapter 3).

Generally speaking, there is only one language, Mandarin, used in Hebei Province, and almost all people are of one ethnicity—Han, only a few are Muslim; therefore, there is almost no cultural diversity in Hebei, China.

In the center of the northern part of Australia lies one self-governing Territory—the NT with a total area of 1,347,525 km². It is 17.5 percent of total territory of Australia. The NT is composed of sea, plain, desert, and wet land. There are two seasons in the Top End tropical area: dry and wet seasons, but in the central area of NT, four seasons (spring, summer, autumn, and winter) exist (Chapter 5).

With its special location, and Australia immigration policy, the NT has great language diversity, and also language is based on culture and tradition; therefore, cultural diversity appears. The 1996 Census shows that 22.5 percent of the Territory's population was born overseas and the Aboriginal population was 28.5 percent of the NT's population and represented 13.4 of total Indigenous population in Australia (Chapter 5).

Generally, Territorians speak English, but French, German, Chinese, Japanese, Italian, Greek, Indonesian, Tagalog (Philippine language), and so on as well as many different Indigenous languages, are spoken by the population.

It appears that there are big differences in the study sites of both countries in terms of geography, language, and culture diversity.

Educational Development

Educational development for both systems is an important issue to be considered and compared in this study.

In Hebei, China: Currently, the general system of formal education comprises four stages: the primary, the junior secondary, the senior secondary, and the higher education. Rural Hebei usually provides only the first three stage as 6-3-3 system (six-year primary, three-year junior secondary, and three-year senior secondary), whereas in the NT almost the same system exist, a 7-3-2 system has been used (seven-year primary, three-year junior secondary, and two-year senior secondary). After the establishment of the People's Republic of China, education in the above four stages has been improved significantly. Generally speaking, there have been five periods of educational development that correspond to the political, social and economic changes, and transformation in China, namely: "1949–57, the transition to socialism"; "1958–60, the great leap forward"; "1961–65, Readjustment and recovery"; "1966–76, cultural revolution"; and "1976–present, post-cultural revolution reforms and opening-up." While in the NT, educational development in this study is from 1978 when NT became a self-governing Territory of Australia.

The fastest and most important stage for educational development in Hebei, China, emerged from the period of post-revolution reforms and opening-up to the outside world, which started from 1976 until now. From that time on, the Chinese Central Government and Hebei provincial government have considered many issues through policy, documents, and laws about educational reform and development, and great efforts have been put into practice. For example, in 1983, the government of Hebei Province initiated the reform of the administrative system of education in rural Hebei and started to transfer the responsibility of primary and junior secondary schools in rural areas to the local government. In 1985 educational reform legislation officially placed rural secondary schools under local administration. In 1985 the definitive reformulation of the earlier decrees came with the "Decision of the

Reform of the Education System." This has been the guiding policy document of reform for all levels of education during the reform and opening-up years. In 1986, the Law on Nine-Year Compulsory Education took effect, and other issues related to technical and vocational education. All of those changes came from the guiding policy documents of educational reform for all kinds and all levels of education and guaranteed education for all in Hebei Province. The main achievements were to popularize primary education, to popularize nine-year compulsory education, to enlarge higher education, as well as to develop technical and vocational education and other kinds of nonformal education (Chapter 3).

In the NT, Australia: The significant educational development in the NT started after the NT became a self-governing territory on the first of July 1978, almost the same time when Hebei developed its educational reform after the "cultural revolution." The changing of educational administrative system from Commonwealth into Territory government has given Territory more motivation, opportunity, and responsibility to make significant improvement for various areas of education in NT. The evidence has shown that a lot of changes have taken place. For example, school design and construction used to copy the southern states. However, after 1980, the construction of schools was more closely suited to local climatic conditions (Chapter 5). Furthermore, Annual Reports of the NT Department of Education from 1982 to 1985 described educational reforms and innovations in many areas, such as curriculum initiatives and policies, commitment to computer education, human resources, and so on.

Unlike Hebei, China, in which there is no Indigenous population, and the people speak and use the same language, the NT includes many culturally and linguistically diverse Indigenous communities with their own culture, languages, and knowledge. As indicated earlier, in the 1996 census, Indigenous population occupied 28.5 percent of the Territory's population, and also that number represented 13.4 percent of total Indigenous population in Australia. In 2001, Indigenous students occupied 38 percent of the NT government's total student population (DEET Annual Report 2001–02, DEET of Northern territory, p. 115); therefore, Indigenous education and development of Indigenous communities became an important component of educational issues in NT. This is a significant departure from the situations in Hebei, China.

The earliest VET in China may be traced back to the industrial education in the 1860s, more than 140 years ago. The main content at that time (late Qing Dynasty) was to study western technology and train manpower with practical skills. Later, in 1917, the "Chinese Vocational Education Society" was established, which was a precedent for the joint provision of vocational education by the educational sectors and industrial sectors. However, the slow economic progress and backward industry hampered the development of VET in China before 1949. After 1949, when the People's Republic of China was founded, some progress was achieved, but unfortunately, the normal pace of VET development in China was seriously affected by the "Cultural Revolution." Chinese VET has achieved tremendous development after 1978 (post "Cultural Revolution and Reform Period"). China is a country in which the government plays a very important role for the main activities of education. Therefore, government develops the relevant policies, documents, regulations, and decisions to ensure the educational activities occur as planned. This kind of intervention also happened in VET. For instance, in 1991, the State Council formulated the "Decision on Energetically Developing Vocational and Technical Education" which identified the tasks and objectives for the further development of the VET in the light of economic and social development in the 1990s in China. The "Outline on Reform and Development of Education in China" drawn up by the CPC Central Committee and the State Council in 1993 required government at various levels to attach great importance to VET, make overall plans, and energetically develop VET. The Outline on Reform aimed at mobilizing the initiatives of all departments, enterprises, institutions, and all quarters of the society to provide VET of multiple forms and various levels. Furthermore, in 1996, the first "Vocational Education Law" in China was formally promulgated and implemented, so as to provide legal protection for the development and perfection of VET (Vocational Education in China, China Ministry of Education, Monograph in Chinese).

The Chinese system of vocational education consists of education in vocational school and vocational training. Vocational education in China is provided at three levels: junior secondary, senior secondary, and tertiary.

When it is conducted mainly in the junior vocational schools, it is aimed at training workers, farmers, and employees in other sectors with basic professional knowledge and certain skills. Thus, junior vocational education refers to the vocational and technical education conducted after primary school education and it is part of the nine-year compulsory education. To meet the local needs of labor power for the development of rural economy, junior vocational schools are mainly located in rural areas.

The senior secondary level mainly refers to the vocational education in the senior high school stage. It is composed of specialized secondary schools, skilled worker schools, and vocational high schools and it is the mainstay of vocational education in China.

Tertiary vocational education programs mainly enroll graduates from regular high schools and secondary vocational schools.

Apart from vocational education, various vocational training schemes exists in China, mostly conducted and managed by the Department of Education and Labour, but enterprises are also encouraged to provide vocational training for its own employees.

In rural China, the vocational education is mainly conducted by the specialized secondary schools on agriculture and forestry, rural vocational high schools, and farmer schools (http://big5.xinhuanet.com/gate/big5/news.xinhuanet.com/zhengfu/2003-02/11/content_723815.htm, accessed on April 6, 2011).

Compared with Australia, the vocational education/training in China is still at a low level in terms of organization, management, curriculum, and other aspects.

Currently, compared with China, Australia has a very organized system of VET in terms of development, management, and promotion of the National Training Framework, national strategy, curriculum design and advice, evaluation, and developing advice to identify and plan for future growth requirements. For example, there is a national training organization (The Australian National Training Authority—ANTA) in charge of above-mentioned activities and objectives.

VET in Australia started in the mid to late 19th century with the establishment of mechanics' institutes, schools of mines, and technical and working men's colleges to develop the skills of Australia's working population, typically men. For more than 100 years, the Australia VET system has continued to respond to industry, individual, and community needs, focusing on capturing the best advice possible from industry; meeting client needs; and aimed for clearer, higher quality standards, all within a nationally consistent, quality VET system.

Vocational education/training in Australia today is "education and training for work." It exists to develop and recognize the competencies or skills of learners. Traditionally, it has been viewed as postsecondary, nonuniversity education and training, focusing on apprenticeships.

But reforms in the past decade now see vocational education and training programs offered in secondary schools, stronger links with university study options, and six levels of qualifications offered in most industries, including high growth, new economy industries. In 2001, nationally, there were over 4,000 registered training organizations, including TAFE institutes, private training and assessment organizations, enterprises, universities, schools, and adult education providers. VET in Australia is an industry-led system, and Australian federal government, state, and territory governments provide the policy and regulatory frameworks for the VET system. Governments implement the National Training Framework (which includes Training Packages and the National Quality Training Framework) to enable consistency, quality, and national recognition of provider services. Governments also provide approximately half the funds for the system, and the other half being provided by enterprises and learners themselves (http://www.anta.gov.au, accessed on April 8, 2011).

In Australia, VET is composed of schools, TAFE, and adult and community education institutions. Therefore, when we talk about VET in Australia, it is hard to ignore TAFE. TAFE is a main body of VET in Australia, and the formal education for many careers. Generally speaking, there are three levels of education in TAFE: the first aims at education to obtain certificates 1–4, and specialize in short courses in business, office secretary, industrial design, house care, tourism, and health sectors; the second is for professional diplomas, which is to meet the both personal career needs and a qualification to enter university, normally such study takes two years; the third focus is on degrees, specialized in applied science and computers, emphasizing on theoretical, rather than practical like the first level. This last level within the TAFE sector is in its infancy. Degrees are typically reserved for universities. However this is changing (http://www.tech.net.cn/y_jyjs/gjgn/au/2567.shtml, accessed on April 8, 2011).

Primary and secondary education for both systems appears to have similarities in terms of schooling years, government responsibility, and significant development in the past 20 years.

China and Australia have almost the same great land areas, which required both governments to find ways to deliver the education into those big areas. The difference is that in China, especially rural areas, people usually lives in a village, but in Australia, people's living is very scattered outside the major urban areas. Sometimes it is more difficult to get people together for training and other educational interventions outside the urban areas.

China has for a long time had centralized, well-established, and very organized government. Education, of course, is a general function of government. Australia is different. The constitution of the Australia which includes federal government and state governments in term of education is quite different to China; education in Australia is a state/territory government responsibility, but sometimes, the federal government intervenes and supplies funds for specific education. For example, if the federal government wants to improve scientific laboratories, it then offers money to improve laboratories. If the territory government is willing to accept the program, then funds are allocated for this purpose. Consequently, the federal government through policy initiatives will supply money and keep actively involved in different education programs. This separation of funding and power is a key different in education between China and Australia (In 19, 2004).

Even though it started early (about 140 years ago), VET in China is still in an initial stage, as traditional educational thought paid less attention to VET. The most important development period of VET is very short, carried out only after 1978.

In Hebei, the acceptance of vocational secondary schools was slow, at least initially. The perception lingered that these educational streams were only for those not able to pass in the traditional stream to climb the social ladder through higher education. In rural China, a senior secondary-school graduate is considered as an educated person, although secondary schools are viewed as a training ground for colleges and universities. And, while secondary students are offered the prospect of higher education, they are also confronted with the fact that university admission is limited. After 1978, the serious problem appeared that a large number of secondary school graduates returned back to their villages without any practical skills and could not meet the needs of rural economic development. Therefore, to develop secondary vocational education then became a major policy of the central government.

Although the NT has a well-established education system and high level of literacy, people in rural and remote and Indigenous areas still face problems.

The previous section has described and compared the general factors of both systems in order to present a whole picture for comparison.

The following section concentrates on the educational comparison of the roles of university for rural development.

Comparison Two: The Roles of University for Rural Development

Universities, especially agricultural universities in developing countries, have clearly identified that rural development is their main mission. Universities transform their knowledge base from research and apply it into rural areas.

It is clear that Chapter 4 has built up the necessary description of AUH and its rural development service to identify that from the beginning of the 1980s, AUH in China has adjusted and reoriented its programs to rural areas as well as played an active and constructive role in rural development. CDU in Australia is a comprehensive university in a developed country. It is in a unique environment. It is the only university in a broader NT, and only university outside Victoria with dual-sectors (higher education and TAFE). CDU has understood the importance of rural development and it has put into practice rural involvement activities through its research, instruction, and consultancy, both in the higher education and TAFE sectors. All such contributions emphasized human resources capacity building and the empowering of local people in rural, remote, and Indigenous communities.

AUH is an agricultural university in a developing country. Its experiences and involvement in rural development service have shown that it has a key role to play in ensuring that critical knowledge and skills are imparted into rural communities to build human resources capacity. The impact of their involvement has made a significant contribution to the quality of education, on the improvement of rural life, and on the sustainable natural resources development. Of course, as an agricultural university, traditionally, its mission focused on crop and animal production. And it is a place for research, teaching, and extension/consultancy. But along with the educational reform and opening door policy in the early 1980s in China, AUH has redirected its mission towards the broader aim of supporting rural development. The successful stories of AUH have proved that if a university, especially an agricultural university in a developing country has identified and redirected rural development as its main mission, then the transforming of knowledge base into rural areas and contribution of human resources capacity building in rural areas could be realized. The important issue found from this comparison is that any university (nonagricultural university in developing country) like CDU, if rural development becomes part of its mission or at least consideration, then the new methods, new teaching and learning models, and new partnerships could be created so that knowledge transformation into rural communities could be realized.

Different Levels of Government Have Committed Strongly to Support Universities to Deliver Rural Development Service Both Financially and Institutionally, and with Relevant Strategies and Policies

Generally speaking, rural development is often seen as a government responsibility. The university, as an educational provider, has as its main mission, teaching, research, and consultancy; therefore, in order to orient the university towards delivering rural development programs that encourage efficient use of the human resources and knowledge, government should have special policies to commit universities towards rural development as one of its services.

In China, government has a strong commitment for AUH to serve rural development programs. For instance, from the initial stage of Taihang Mountain development project to the follow up and other programs, different levels of government encouraged and supported AUH financially, institutionally and with policies, strategies, and government policy documents. All these commitments ensure that AUH can successfully deliver its rural development programs and achieve the expected outcomes. It is also clear that China is a developing country which has had a centralized government for a long time; government played a major role in many aspects of project development, without government support, coordination, management, and commitment, the university itself would find it very difficult to undertake some rural development programs.

Apart from financial support, the government's commitment for CDU to undertake the rural service program is mainly built on establishing partnerships or cooperative activities. Government priorities, for example, were given to those programs which concentrated on areas in rural, remote and Indigenous education. The NT government, for example, supported the amalgamation of the CDU and Centralian College, in which CDU became the largest public providers of TAFE in the Territory, improving its ability to cooperate with secondary schools and also increasing its presence in Central Australia and research in fields of desert knowledge. One of the principles of the NT government's partnership with CDU is to set up "particular projects enabling Indigenous social and economic development" (Higher Education Review-Submission to Collaboration Task Force, Office of Territory Development, NT, September 2003, p. 2) in the territory.

The government commitment for both universities to serve rural communities has similarities in many aspects, and thus the key difference that emerges is that AUH is an agricultural university in a developing country, focused on agricultural sector, and government in China has played a strong role in rural development activities, whereas CDU is a university in a developed country, focused on comprehensive fields, and government's intervention for rural development program is far less. Furthermore, AUH in China is mandated by the state to serve the agricultural and rural needs of the province (Chapter 4), whereas in CDU, Australia, government cannot do the same thing, as Australia is a democratic government system (In 22, 2004).

Consequently, it can be concluded that different levels of government involvement is a key role for the successful transforming of research, knowledge, information, and skills from universities into rural communities for their development, either in developing countries or in developed countries.

Communities Have Paid Great Attention to the Programs Carried Out by University for Rural, Remote and Indigenous Areas Development

The successful rural development programs carried out by AUH and CDU depend in part upon the positive participation of the communities in different ways and thus this is a common requirement in both systems. Concerning participation, people should have enthusiasm, motivation, and be interested in and involved with the programs. For instance, some rural development programs achieve their common goal of economic development; some are intended to accomplish social and educational development; some may want to build their own individual personal capacity and empower themselves and some have other initiatives. "Ensuring community participation depends upon factors such as culture, capacity to respond, awareness of the issues, and commitment to deal with the problems" (UNESCO, 2001). In AUH, most programs are concerned with agricultural extension since Hebei is an agricultural province. A large rural population and a limited amount of cultivated land require more and more useful techniques and active skills to be put into practice to increase crops and animal production and to improve people's living standard. People themselves also use education and agricultural extension, as a bridge, to better themselves as governments continue to have a growing emphasis on human capacity development and economic development. Community participation on AUH's rural development programs results in "mobilizing resources, sharing responsibilities and establishing a sense of ownership to sustain community development activities" (UNESCO, 2001). The very important issue of community participation in rural development must be considered if people are to be mobilized properly as is highlighted in some stories on AUH identified successes (Chapter 7).

CDU's emphasis on participation has links with a community development tradition. Communities themselves want to be independent and prefer to engage the people with knowledge and skills in their communities rather than someone from outside. This self help often motivates the community. Aboriginal communities often want the people with knowledge and skills in their communities to maintain their knowledge. The fear is if they come to study in CDU or other universities, the community will lose them when they graduate (In 21, 2004).

It is a basic requirement for development that the community itself is willing to accept and participate in the university's activities. For universities to serve rural, remote, and Indigenous communities all programs to fulfill their objectives must have community participation. In this regard, AUH and CDU have found out that the successful way to maintain community awareness and mobilize the community into action is to have the community participate early in the identification of the aims of the program.

Summary

First, university itself must redirect its mission and put rural development as its main priority; second, universities' rural development programs should be undertaken under the coordination, leadership, support, and commitment of government at different levels; third, the expected outcomes and successful implementation of universities' rural development programs cannot be realized without community participation. It is apparent that two universities have satisfied many aspects of these three issues. AUH has more interventions in this regard, the evidence shows that it is an agricultural university in a developing country, and that Hebei Province is an agricultural-oriented province, and that most of population in this province lives in rural areas. It is also considered that there are some weaknesses in CDU's rural service intervention because there is less rural population, which is living spread over a vast area. There is also difficulty associated with specific environment and cultural diversity.

Comparison Three: Effective Model for Transforming Knowledge, Technology, Information and Skills from University to Rural Communities (What Activities should Be Used by Universities to Carry Out Rural Transformation Development Programs)

There is no doubt that a university can make a positive and effective contribution for rural development through its knowledge, information, and skills transformation. The considerations that need to be analyzed in the sector are: How to realize the transformation? What models are used by universities to do so?

Establishing Demonstration Communities

As an agricultural university, AUH believed that specific demonstration is an effective way to transform knowledge, technology, and skills from its research into rural communities. Consequently, AUH has set up many technology and actual skill demonstration bases in different countries, townships, and villages. In some cases even a few demonstration households in one village have been set up to show the people in such surroundings the importance and benefits of technologies and skills. Many successful cases in previous chapters indicate that this is an effective way to efficiently transform knowledge, technologies, and skills into rural communities. Generally, there are three stages for AUH to do so: first, at the beginning, farmers used to refuse to accept new techniques due to their low educational level, low awareness of the importance of techniques, and low ability to bear the risk of changing traditional methods since they were not sure whether or not they would be any benefits. In this case, the staff from AUH introduced very simple, easy-to-learn techniques with less investment and more yields. These demonstrations showed that the farmer could get out of poverty sooner. This helped increase more motivation to learn techniques and skills.

Second, model households were set up as the effective method used by AUH. Farmers in poverty-stricken areas usually respond to new techniques overcautiously and take a "wait and see" attitude. They will refuse to adopt techniques without visible evidence. To them "seeing is believing." They want to know if a new idea or technique works or not. Due to this reality, staff usually focused on better-educated farmers, and some key persons, especially members of the village committee. When other farmers saw the successful results, which had helped these pilot-farmers out of poverty, they immediately changed their attitude. Thus often an attitudinal change is necessary to get rural communities to accept innovation.

Third, establishment of demonstration communities. After initial success of the demonstrations, the project is expanded further, focused on enlarging the demonstration area in a certain selected community, where many new techniques and research findings suitable to local environment could be examined and demonstrated. Farmers can make their own choices after comparing. Before and after demonstration trials, the university staff could also conduct applied research to solve new problems that appeared in agricultural production while the community adopted the transforming techniques. The demonstration bases serve as technology-dissemination centers. They could also serve as on-site instruction farms where university students could come into contact with farming practices and find out the actual needs of farmers. Therefore, these "integrated teaching, research, and agricultural extension" into demonstration communities hold the key to the successful intervention in the Taihang Mountain development model of AUH (FI8, 2004).

This is no evidence to show that this approach happened in rural NT, or that a similar method was used by CDU through this study.

AUH is an agricultural university in a developing country. Extension is one of its main missions. And also in Taihang Mountain areas and other poor areas in rural Hebei, farmers' educational level is very low. "The demonstration is a powerful method used to help farmers especially those who are not good at reading" (Cakley and Garforth, 1985, p. 49). Therefore, the agricultural technology demonstration carried out by AUH has given farmers the opportunity to observe the difference between a recommended new technique and traditional one, between poor seeds and recommended seeds. Some other innovative methods have also been used by AUH. Within the NT, the milieu is different. There is less agricultural activity in a broad NT, less population living in rural NT area, and a limited use of demonstration as a policy although there is some evidence that visiting teams of academics use the techniques as a teaching tool.

University's Professors and Staff are Willing to Go Out of the Campus and Spend Time and Live in Rural Community

Comparison: It is clear that a good and effective way to transform knowledge and skills from university to rural areas is that the university staff (lecturers, professors) go out of the main campus into the communities, living with community members, finding out what are the real needs of rural people and communities, and as a consequence to help them to build up those knowledge and skills. In AUH. as an agricultural university, rural development and agricultural extension are the main public mission statements. University staff's daily work through their teaching and research activities thus focus on such mission. In this case, professor and staff often visit rural communities and spend time with rural people. This is a normal practice of any agricultural university. The key difference between AUH and other agricultural universities is that AUH has put rural development as its priority. There is a long-term target for comprehensive development in rural Hebei through continuous integration of teaching, research, and agricultural extension, as well as the combining of agriculture, science, and education, and the integration of literacy education and technological extension with agricultural production.

More than 20 years of practice has shown that staffs, especially professors, from AUH have paid their individual visits to villages. They carry out face-to-face training and demonstration in rural areas and they are more acceptable to the community, because rural farmers often trust professors. The farmers are more likely to listen to the instruction and advice given by them and they will be more grateful for this individual attention. Oakley and Garforth (1985) also shared the same opinion: "Personal influence of the extension worker can be a critical factor in helping a farmer through difficult decisions, and can also be instrumental in getting the farmer to participate in extension activities."

Professors and staff of CDU, who visited research sites and rural communities, maintain that such visits to rural Indigenous communities are a most effective method of getting information across.

It can be concluded that both universities send professors and staff to rural communities to carry out development programs, but it seems that AUH has paid more attention in this regard in terms of numbers of visiting staff, visiting times, and spending period in rural areas because of its specific situation as an agricultural university.

University Has Set Up a Network and Built Up a Partnership with Relevant Institutions and Organizations to Share Resources so as to Transform Knowledge, Technologies, and Skills into Rural Communities

AUH is the only agricultural university in Hebei Province. It has limited human and financial resources available, so it can only offer a limited direct support and service to rural communities. But the research achievements on crop production and animal husbandry as well as in other fields related to rural development have remained at high levels. In order to fully use research results and resources of the university and transform them for the benefit of the rural communities, network and partnership building became important.

In Hebei Province, there are 139 counties (or cities), each of them has a vocational and technical center; furthermore, in each county, city, or township, adult school, and farmers' night school have existed. Most of these are located in rural areas or somewhere close to the rural areas with agriculture and related science and practice as their disciplines. Most of the students are from rural areas and looking for the practical skills necessary for the development of the rural economy instead of textbook knowledge. Since such educational recourses are located in every county/city, and cover most of the local conditions, AUH set up a network throughout these areas and built partnerships with those centers and schools to enlarge its rural service capacity (Appendix 2). AHU helped such centers in teacher training, curriculum development, school management, program guidance and consultancy, and so on.

Apart from CDU's campuses at Darwin, Palmerston and Alice Springs, there are centers located at Katherine (NT Rural College) and a few regional study centers located at Jabiru, Nhulunbuy, Katherine, and Tennant Creek. These locations allow the CDU to have a spread of educational resources and facilities across the breadth of the Territory with exposure to tropical and desert environments as well as the rich indigenous culture of Australia. Each regional center acts as CDU's link to regional and remote NT, making courses and training accessible to more Territorians. This network system seems to be reasonable in terms of the serving areas, local conditions, population distribution, and cultural diversity.

As universities, both CDU and AUH have limited resources in terms of human and finance. They thus can provide some support to rural, remote, and Indigenous communities. It is difficult to cover broad areas in rural Hebei and rural NT. The practices from both universities of network establishing and partnership building become an effective way for their intervention in rural communities. As an agricultural university and serving a considerable population in rural Hebei, the network of AUH appears more complete, creative, efficient, and considerably larger in numbers, whereas in CDU even the network is small in number, but the function of enlarging its service to a broad areas and people, and meeting the people's learning needs in different environment and different conditions has been realized.

University's Training, Research, and Extension Programs Have Been Closely Linked with the Local Needs and University Has Considered Any Specific Conditions and Situation in the Program Target Areas

AUH's rural development programs including training, teaching, and extension have close links with local communities taking consideration of local conditions. These two principles (local community and local conditions) have been included in the program design and implementation. For example, from the beginning of 1980s when the Taihang Mountain Development program was carried out, local conditions and local needs were two important issues for the projects to be considered before the implementation. From that time on, professors and staff spent a long time in this mountainous area and made comprehensive system surveys and analysis on the existing problems and potentialities. Finally, "Resources Investigation Report in Taihang Mountain Area," which included 15 specialized areas was developed as a resource for future work. All this first-hand data and information was used later by project implementers during development.

Other micro case studies have also shown that local conditions and local needs both for communities and for farmers have been considered. These include the development of new seeds varieties and animal breeds in practical ongoing courses and research investigation of students and staff.

Paying continual attention to the real needs of rural, remote, and Indigenous communities is also the consideration of CDU. The examples are found in the teaching, research, and extension activities of CDU Rural College, Tropical Savannas CRC, and Center for Indigenous Natural and Cultural Recourses Management.

Both universities have the same consideration on the rural development programs, which focused on their training, research, and extension activities with a close link to the local needs and the local conditions. Those kinds of development programs for AUH include training, research, and extension, whereas CDU has much more concentration on training and research, although TAFE extension programs are developed and implemented.

Apart from the University's Contribution for Rural Development, the University Itself Has Also Grown While It Serves Rural Development

The responsibility of community development in developing countries such as China often falls to the university, especially agricultural universities. AUH has made various efforts to do this for more than 20 years. During these rural practice and rural service, AUH itself has also been promoted and upgraded in the aspects mentioned below.

The university scale enlarged and student enrolment numbers increased. Since 1979, the construction of AUH has been carried out faster. In 1995, a new AUH emerged from the merging of AUH with the Hebei Forestry College. The size of AUH has been increased in terms of faculties, departments, and number of staff and students. For example, in 1980, AUH operated 7 departments and 10 specialties. In 2001 AUH covered 25 colleges and offering a total of 51 bachelor degree courses, 24 masters degree courses, and 4 doctorate degree courses. The total number of student enrolment is more than 20,000 where more than 800 postgraduates were for master and doctorate degrees. The teachers and staff members' population is still 2000, the same as that of previous years (*University Information Handbook*, 2001, p. 3).

The specialties were adjusted and developed along with the adjustment of production requirement. Productivity has developed constantly and the structure of agriculture also changed. These changes called for the corresponding change of specialization offering of agricultural universities because the specialization framework and structure radically reflects the level of social economy and technology support. In order to fit in with the changes of rural industry structure, especially to meet the urgent need of rural commodity production and market economy development, AUH broke away from the traditional concept of agricultural production when setting up course specializations. The different fields of specialization were adjusted in accordance with the development of comprehensive and general agriculture. The social demand changes quickly, especially under the market economy condition. AUH took the attitude of respecting reality and being practical and realistic when putting up fields of specialization. Present needs were considered as well as long-term viability so that the university gained a stable and healthy development.

Research deepened and developed further and integrated with teaching/training and extension. A number of research programs had come in after the research project on developing the Taihang Mountain in 1979. Some of these projects had gained attention on the provincial and national levels. Some international institutions like UNDP have recognized some research outputs by the university. Currently several research linkages have been fostered with other universities around the world. Research programs, research equipment, and instruments increased, and funds sources for research and extension broadened.

Students' awareness of science and culture improved and their capabilities enhanced. Students can not only improve their practical techniques, perceptual knowledge, and understanding of the society to enhance their comprehensive quality, but also shorten their time for adaptation and strengthen their competitive power in employment after graduation. By launching the social practice activities (the student social practice activity is referred to that during school holidays, students organized by school visit factories, villages, communities, hospitals, and other areas to have investigation in order to understand them, as well as make the contribution to them by their knowledge), the students' spirit for the love of the land and to be of service and become involve in the development process is enhanced. Giving them firsthand experience, helped to develop talent, thus allowing more creativity to surface and enhancing more innovative approaches. From theory to practice and from practice more concepts can be derived creating possibilities for the development of new theories.

Similarly, CDU emerged out of Darwin Institute of Technology (DIT), and Northern Territory University (NTU). In 2004, NTU (former name of CDU) has merged with Alice Springs' Centralian College joined with NTU to become CDU to ensure that it was a University for all Territorians, no matter where they lived (http://www.cdu.edu.au/newsroom/stories/2004/april/healice/, accessed on August 13, 2011). CDU grew from a community college, largely funded by federal and Territory governments, to a full dual propose (Higher Education and TAFE) university funded under the national university system. This enlargement has given CDU an opportunity for more people to be educated and trained as well as to be more involved in rural community services. Northern Territory Employment, Education and Training Minister Syd Stirling said that:

> One of the main aims with the creation of Charles Darwin was to ensure that it was a University for all Territorians, no matter where they lived. This investment in Central Australia will help to achieve that vision. Importantly, it means more Central Australian students will be able to pursue their studies locally, rather than having to move to Darwin or interstate and staying there. (http://www.cdu.edu.au/newsroom/stories/2004/april/healice/, accessed on August 13, 2011)

The practices of both systems have shown that universities can also be promoted and developed and make benefits for themselves while they serve rural development. This could become a critical index of deciding if the university will volunteer to become involved in rural reconstruction under a market economy.

Summary

This section has examined and compared effective models for transforming knowledge, technology, information, and skills from university to rural communities. It concludes that AUH has satisfied most conditions, but some weakness to meet the condition appears in CDU, since there are differences of development level, location, population, learners' needs, and less reliance on agriculture. The time and dimension and government intrusion are also factors that should be considered (e.g., planned economy for Hebei and a market economy for NT).

Comparison Four: Innovative Approaches Undertaken by University for Rural Development (How Are These Models Realized in Practice)

Digital technology, Internet access, and other simple and effective media have been used by the University Rural Education and agricultural extension. Digital infrastructure has been extended from urban to rural areas. Then efficiency in terms of cost, staff's time, and learner's achievement will be much more increased.

The simple and effective media used by AUH to deliver information to farmers referred to broadcast and TV. And in some areas, for example, plant protection, a hotline telephone was used by professors in relevant careers. Some weaknesses existed in rural Hebei with regard to this infrastructure. Agricultural information sometimes is inaccessible to the users at the grass roots level; incomplete coverage of agricultural information distribution often occurred at the grass roots level, which constrains information transmission. Only recently have Internet access and other digital services become available for agricultural information and rural development service. With the rapid development and application of information technology, geographic information systems, the system of remote sensing information, the system of global positioning, the expert systems in agriculture, horticulture and animal sciences, as well as the agricultural analogue techniques have become available for research, training, and rural development practice.

In CDU, digital technology and internet access have been widely used for external learning and training workshops in rural, remote, and Indigenous communities. The relevant infrastructure and many facilities necessary for distance training and learning are also available. Connections between campuses and regional centers have been established. All those digital media have made the rural development training and other activities more efficient, less cost and more time saving. The modern technologies of geographic information system, remote sensing information, global positioning system, and other High-Tech media played important roles in rural research and development programs at CDU. Besides, audio-conference, video-conference, and tele-centers make online conversation and online learning available off campus. For this to be totally effective rural areas must be able to access it. Currently, government policy (Telstra) continually aims to improve this access with the national roll out of broad band technology.

Today news media tend to merge together to transmit information and deliver knowledge. CDU has made a full use of modern media for its all interventions with rural, remote, and Indigenous communities, which has shown to be more effective and efficient than the previous print-only courses. In AUH, the situation is different. Even though some modern technologies are available, the traditional media still lies in a mainstream of knowledge transforming, since a considerable amount of rural population has less development in IT infrastructure in rural areas and people's awareness to use them.

Organizing Community Members into Various Technical or Learning Societies, Associations, or Other NGOs under the Guidance of University Staff. Then a Learning Society Could Be Created

It is a common understanding that only when farmers' knowledge level and their ability to accept science and technology and their enthusiasm to learn new techniques and skills has been stimulated, can they possess the ability to develop themselves. Using the motto: "to deal with ignorance before developing mountains," AUH helped to establish various farmers' technical associations using the principle of, "Experts taking a leading position, with local government coordination and model households selected as the core; farmers will join in the Associations by themselves adopting agricultural practical techniques as the top priority."

This approach results in the process of farmers' positive learning instead of being trained passively.

Based on this principle, AUH has successfully organized more than 10 different farmers' technical Associations, which have brought about obvious social and economic benefits, for instance, "Mushroom Association" in Tangxian County, "Chicken Association" in Laiyuan County, "Red Fuji Apple Development Association," "Watermelon Association," and "Peach Association" in Shunping County, and the similar organizations of apple, vegetable, and maize in Wuyi and Zanhuang Counties. Currently, the Association's functions are composed of training, research, extension, production, and marketing. This approach allows for the Association to become a learning, research, and production society.

These Associations have played a very important role in promoting the development of rural education and production. They are helpful for farmers to learn techniques and skills and apply them positively. Such Associations also allow for the acceleration and extension of scientific research findings into rural areas to improve the farmers' awareness of science and technologies. An example of this approach is found in the Red Fuji Apple Development Association of Beicheng Town in Shunping County set up by Mr Huangpu Zhongsi, an associate professor in the Horticulture Department of AUH in November 1990. This Association has brought about considerable economical and social benefits. It not only provides technical training courses, but also instructional services as well as farm inputs and marketing of products. The Association is mainly composed of model households while Mr Huangpu Zhongsi acts as the technical consultant. Every year, Professor Huangpu goes often to the rural areas to hold technical training classes. Each lasts 1–3 days according to the farmers' practical needs in production. The curriculum include: the management of orchards, the prevention and control of apple trees' diseases and elimination of pests, management of water and fertilizer, the storage and postharvest handling of fruits, management of seedlings, as well as the establishment of orchards, and so on. Each time over 1,000 farmers are trained. In 2001, the Association has been extended to more than 10,000 memberships, having Shunping County as the center, including 110 villages in 6 counties nearby. The area of total orchards is up to more than 2,000 ha. The apple output produced by the association members has been up to 37,500 kg per ha, with the total output of 30 million kilograms. Some production has even exceeded that of Japan, the original place of Red Fuji, in terms of the yield and quality of apples. The Red Fuji apple production in Shunping County has become a primary industry in Taihang mountainous region in Hebei Province.

Similar societies in Australia and NT from farmers themselves and government initiatives could be found, like the Dairy Farmer Cooperative, Goat Breeding Society, Pasture Protection Boards, Water Resources Commission, etc. There is also a great deal of experimentation in the NT in agricultural output, especially with regard to rice and mango production and other tropical produce. Much of this analysis is driven by local entrepreneurs and government departments of agriculture and forestry, etc. The university via Katherine Rural College also contributes training and research to this output. Many functions of these associations are done in NT by various Government Boards, for example, Pasture Protection Board, or Farmer Associations and Cooperation set up by farmers themselves.

Another case to show that the CDU's intervention to set up and expand a learning society could be found out from a innovative e-learning project set up in Katherine and Alice Springs, which aimed "to enhance learning and teaching outcomes in remote areas" (http://www. cdu.edu.au/newsroom/stories/2004/september/elearning/index.html, accessed on September 6, 2011). They use satellite as a media and IDL as a technology to transmit knowledge delivered by lecturers such as showing video footage, live demonstrations of materials, and other electronic content like PowerPoint files, which students can view on the computer screen. And also "the IDL sessions allow time for the students to gain valuable knowledge by interacting directly with the lecturer and discussing their studies" (http://www.cdu.edu.au/newsroom/stories/2004/september/elearning/index.html, accessed on September 6, 2011).

The innovative way is to involve students from remote areas into a similar teaching and learning classroom as inside of campus, and create "the opportunity to talk face-to-face, demonstrate learning concepts and receive direct feedback from students is so important to the learning experience. I see real potential in the application to training, particularly for students who are isolated" (http://www.cdu.edu.au/newsroom/stories/2004/september/elearning/index.html, accessed on September 6, 2011).

It is apparent that learners' or farmers' associations can overcome the weakness of the learning in rural areas with a broader rural coverage, and insufficient of teaching resources. It has also been shown that training core farmers, organizing associations in the rural communities, and farmers learning by themselves under the necessary and effective guidance are acceptable and most welcome by farmers and that this approach will make substantial contribution to rural communities. It is also clear that AUH has done a lot of contributions in this area. Compared to AUH, in Australia, much of these functions are farmer driven, compared to university driven.

The University–Community Partnership Has Been Established Using Package Contract Approach and Establishing Joint Ventures

AUH initiated this new extension service in a range of counties. This approach brought the change of the delivery pattern of technology from only a special department towards an over-all operation pattern by mobilizing many departments like administration, material, supply, financial, and monetary bringing service closer to the needs of rural economic development. The pre-requisite of this approach is to provide services for farmers. The operation of this particular approach is that the service provider, for example, AUH, signs a contract package of technical service. The provider charges some fees for the overall service as the resource is from technical research. AUH and Ding Xing County formally signed the contract for an agricultural comprehensive technological package service in January 1989.

On the basis of self-willingness and mutual benefit, AUH established more joint ventures of teaching, scientific research, and social practice with some bases. By signing contract with these bases AUH gradually changed the extension service mode from totally free to the combination of free and charged services. This new approach is aimed to benefit both sides, to mobilize the initiatives of both providers and recipients and to further enhance the enthusiasm of providers. Through many years of practice, AUH and the local partners of joint ventures have expanded the practice to a bigger scale. The service range was enlarged from science and engineering to art and soft science, from introduction to the expansion of extension service from techniques of increasing production to postharvest processing technology, and from economic development to the combination of economic and education reforms. AUH formed such joint ventures with Shunping, Fuping, and Xiongxian Counties from 1982.

The university–community partnership has also been considered by CDU as a meaningful measure to delivering its knowledge into communities. For example, in July 2003, CDU signed a partnership agreement with NT Government on Internet-based education for remote communities and a virtual DNA facility. CDU's former Vice-Chancellor Professor Ken McKinnon said the partnership agreement signaled a new level of interaction between the university and its community. "As a University for the Territory, it is critical that our intellectual resources, in collaboration with those of the Government, are brought to bear on the issues of most importance to the Territory." This Agreement includes 25 schedules based around four themes, which are: increasing resident professional capacity to address Territory opportunities; meeting government needs; reorganizing the university to better meet Territory needs; enabling Indigenous social and economic development. The main activities focused on remote communities with the specific needs of the Territory, like community development, conservation biology, natural resource management and tropical environmental science, health and diagnostics, as well as Indigenous social and economic development (http://www.CDU.edu.au/newsroom/stories/2003/july/partnership/index.html, accessed on July 1, 2011). CDU also initiated a project at Alice Springs and Katherine for e-learning in collaboration with the Northern Territory Department of Education, Employment and Training who provided funding to CDU. The new practices in flexible learning programs with a focus on interactive distance learning (IDL) technology "allow the lecturer to be viewed by the students as well as showing video footage, live demonstrations of materials and other electronic content like PowerPoint file" (http://www.cdu.edu.au/newsroom/stories/2004/september/elearning/index.html, accessed on September 6, 2011) in remote areas.

It can be concluded that in order to use the resources of AUH, and to have these recourses utilized for the mutual benefits of the university and communities, AUH initiated the university and community linkage and partnership in a broad area. Whilst the approach in the NT was through farmers' organization themselves and government sponsorship recent policy shifts at government level and CDU management are refocusing CDU towards government initiative and community development.

Encouraging University Student Volunteers, Especially Agricultural University Students, to Launch Social Practice Work and Other Practical Courses in Rural Areas

It is a general practice for the Chinese universities to organize students to participate in social practice and other practical activities. In some cases, it is also part of a course requirement. AUH encouraged senior students to launch social practical work at the agricultural interface, and to serve the farmers. They use a one-to-one method, which means one student assists one farmer, and one group is assigned in one village. Activities used were broadcasting, blackboard, bulletin board announcements, farmers' night school learning. The course content for farmers includes promoting new high-quality products, disseminating new technologies, offering training courses for local agricultural technicians, and delivering technical consultancy and on-site instruction. The activities can be undertaken, sometimes in vacation time and students are volunteers; sometimes its programs are carried out during school time as part of course requirement. During this practice, the students not only make a contribution to rural communities, but also experience the value of their knowledge and find out the real needs of farmers. All of these activities are more helpful for their own career development in the future.

Student practical experience is not a university's responsibility or duty in Australia. However, the Student Union provides (through university funding) advice and a range of facilities for students. The university does provide placements for its trainees, for example, teaching and child care, nursing and to a limited extend farming enterprises often such placements can be undertaken in remote Indigenous communities. This focus on rural development, however, is not a priority. However, with the appointment of a professor in rural education with a brief for community development and participation and a professor for Indigenous studies, this is likely to change. In some courses, like engineering, students need to find their own way to look for the practical sites. But in some cases, like TAFE, the opportunities for practice are also given by CDU.

Students participating in practice to serve the communities have a mutual benefit for students themselves and for the communities. AUH has promoted this kind of activities. This has been less evident in CDU in this regard, but changes are happening as CDU reposition itself with regard to the community.

Setting up Rural, Regional, or Night Training, and Consulting Centers

Rural learners or farmers are adults; therefore, the establishment of various kinds of schools for them to meet their learning needs during their spare time from agricultural production or at night is one of the methods used by AUH for its rural adult training and extension intervention. For example, AUH helped establish adult schools in Shuping County. In these schools, there is one infrastructure with multipurposes, which means, there are combined functions of night school, library, information station, plant and animal clinic, agricultural technology extension service station, develop of and new agricultural varieties, machines and fertilizers, chemicals sales are all combined into one place to serve the local communities so as to realize the statement of "diverting science and technology water; through the education channel; to irrigate agricultural farm" (Li, 2000). Professors and staff from AUH are involved in the activities of those comprehensive schools in various ways showing a guidance, consultancy, and even direct services.

In Australia many of these functions are absorbed by agricultural shows in regional communities. In the NT, the Freds Pass, Darwin, Katherine, Alice Springs shows are often one-stop shops, but this response is limited to once a year. The Business School has a Crocodile Firm virtual business. However, university involvement as a policy issue in adult rural training is minimal compared to AUH. With the appointment of a professor in rural education, which briefs for community development and participation and a professor in Indigenous studies, this is likely to change.

In developing countries like China, rural problems appear in the areas, such as a big rural population, much reliance on agriculture, low educational level, less resources both in basic school infrastructure and human resources, and less economic development. The development of those kinds of comprehensive schools in rural communities has proved to be very effective in carrying out and managing, acceptable rural education in rural communities. Such approaches are most welcome by rural people. Some efforts have been made by CDU through its practical Crocodile Firm, and a new professorship appointed.

Training and Encouraging a Large Group of Community Members to Become Backbone Members of Extension Work Force

Generally speaking, farmers in China are afraid to take risks and lack entrepreneurial skills. They are very conservative people living in a close system. Often they resist change and they are unwilling to adopt new techniques or new methods easily without any concrete results. Apart from this, the common feature of farming production confirms that any technical result requires a long time for implementation. The farmers in the extension areas slowly adopt new techniques. The AUH technical team established fixed bases, selected a few community members, normally they are educated people, some community leaders or some people who are willing to participate and have the ability to do so, and trained them to become the farmer technicians or so-called "well-to-do households" family.

Apart from training local key members in the communities, AUH cooperates with government to deliver a program, called "program of one village, one graduate student from AUH." The government provides scholarships and signs contracts with individual students from rural communities to make sure after they graduate from AHU, they will return to their home town. This practice shows that those graduates who have obtained advanced knowledge and are also familiar with local communities' environment have soon become the chief members in their own communities.

All these key members in the communities actively involve themselves in agricultural extension work and demonstration practice. After they achieved economic benefit, they will become the "models" and be followed by other farmers' investigation.

There is a similar problem in the NT where Indigenous students are selected from and by the community to be trained, but, on receiving training, they may not wish to return to the community. There is a Tertiary institution in the NT called BIITE that trains community members. There are two ways for them to be used not only to delivering the training courses, but also make sure when the trainers finish their training courses, they will be back to their own communities. One is to carry out the training courses inside the communities; the other is to select key persons or backbones from the communities (In 24, 2004).

Role models in rural community both in Australia and in Chinese context have proved to be effective in specific rural conditions. This kind of practice has also been considered as one of the innovative methods used by AUH and also the BIITE to transform knowledge and technology into rural communities.

Conclusion

This chapter has examined and compared the two systems, Hebei China and NT Australia as well as CDU and AHU. It is clear that the study has gone into different issues related to both sides, namely:

1. The roles of university for rural development.
2. Effective model for transforming knowledge, technology, information, and skills from university to rural communities.
3. Innovative approaches undertaken by university for rural development.

9
Conclusion

Introduction

Two case studies have been developed in the previous chapters on the roles of university for rural development. A comparative analysis and discussion of the two cases has also been undertaken to comply with Comparative Education Methodology. Some critical issues relating to the study should now be identified and summarized in this chapter, as well as the main findings and conclusion.

Overview of the Book

Generally speaking, the research is intended to find out in depth the relationships if any between: (a) poverty and development; (b) education and development; (c) university and development; and (d) some findings and conclusion related to that development. Therefore, "development through education" becomes an important issue woven through the research study. How to transform the knowledge base from universities into rural, remote, and Indigenous communities to make benefits for the people and the communities becomes a major focus of the "development."

Rural Development

Rural development appear many times in this book, but what is "rural development"? What context is it dealt with? Can it be realized?

Poverty

Before examining rural development, some understanding of poverty and especially rural poverty is necessary.

Even in the 21st century, poverty is still a critical issue faced by many countries, especially developing countries. The general comparative data from the study showed some of serious poverty issues worldwide as follows:

> 3 billion of the world's people (one-half) live in "poverty" (living on less than $2 per day). 1.3 billion people live in "absolute" or "extreme poverty" (living on less than $1 per day). 275 million children never attend or complete primary school education. 870 million of the world's adults are illiterate. Over 100 million people live in slums. An estimated 25 to 50 percent of urban inhabitants in poor, developing countries live in impoverished slums and squatter settlements. (http://www.worldrevolution.org/projects/globalissuesoverview/overview2/briefdevelopment.htm, accessed on June 16, 2004)

Those people face serious situations in terms of basic requirements, such as food, shelters, water, education, health, economy, and other basic human rights.

> Poor people live without fundamental freedoms of action and choice that the better-off take for granted. They often lack adequate food and shelter, education and health, deprivations that keep them from leading the kind of life that everyone values. They also face extreme vulnerability to ill health, economic dislocation, and natural disasters. And they are often exposed to ill treatment by institutions of the state and society and are powerless to influence key decisions affecting their lives. These are all dimensions of poverty. (World Development Report, 2000/2001, p. 1)

United Nations Economic and Social Council gave a definition about what is poverty in its final report on human rights and extreme poverty:

> What is poverty, legally speaking, but a string of misfortunes: poor living conditions, unhealthy housing, homelessness, failure-often-to appear on the welfare rolls, unemployment, ill-health, inadequate education,

marginalization, and an inability to enter into the life of society and assume responsibilities? The distinguishing feature is that these deprivations— hunger, overcrowding, disease, and illiteracy—are cumulative, each of them exacerbating the others to form a horizontal vicious circle of abject poverty. (United Nations Economic and Social Council, 1996, p. 3)

Obviously, poverty is hunger, and lack of shelter; poverty is being sick and not being able to see a doctor; poverty is being without knowledge, and not being able to go to school; poverty is not having a job, and is fear for the future; poverty is unclean drinking water; poverty is powerlessness, and lack of representation and freedom.

Why Poverty

There are different opinions about why poverty exists. And the causes of poverty and its characteristics differ from country to country, and vary from region to region. Some people see poverty as only the lack of substantial materials, they argue that governments and society should provide more; others believe the poverty results from being short of ability or capacity (http://www.edu.an/20030115/3075978.shtml, accessed on June 16, 2011, in Chinese). The Nobel Laureate (1998) Dr Amartya Sen in his two books, *Poverty and Famines: An Essay on Entitlement and Deprivation* and *Development as Freedom* described the reason for the existence of poverty. He highlighted that poverty is not only insufficiency of providers and less income; it is because human basic entitlement have been lost or deprivation has occurred. These basic entitlements include human capacity, education, health, and others (Sen, 1981).

Amartya Sen divided poverty into three categories: income poverty, which means not sufficient income to sustain the basic needs; human poverty, which is reflected in illiteracy, malnutrition, short life expectancy, health of pregnant women, infant mortality rate, etc.; and knowledge poverty, which is concerned not only about the people who are less educated, but also those with less abilities or the inability to obtain, absorb and use the knowledge. This deprivation of capacities or ways to get knowledge often results in poverty (http://www.edu. an/20030115/3075978.shtml, accessed on June 16, 2011, in Chinese).

Even with economic development and social progress, some of the members in society remain in the situation of poverty, since some factors restrict their development, such as the lack of human resources, low educational, and health levels, as well as the loss or deprivation of income-generation capacity and opportunities.

It can be concluded that poverty is a complex issue. It has many dimensions, and it must be analyzed through a variety of indicators: such as, levels of income and consumption, social indicators, and then increasingly indicators of vulnerability to risks and of socio/political access. Furthermore, poverty has many faces, changing from country to country, place to place, people to people, and time to time, and has been explained in many ways. In order to simplify the study, the definition adopted on poverty relies only on income level as the basic measure and the traditional definition of poverty, therefore the analysis can be narrowed down by this restriction.

Poverty in Rural, Mountainous Regions of Hebei

For the population living in rural and mountainous areas of Hebei, China, the literacy rate is not low, according to statistical data in 1999, the literacy rate has reached up to 95.87 percent (Chapter 4), but the problem is that the basic knowledge on reading, writing, and numeracy skills are not enough for the rural population to have poverty alleviation and income-generation. Other factors need to be considered, such as, attitudes and values as well as knowledge and skills related to vocations, incoming-generation, management and social, and political and cultural life. This conception is much deeper than the so-called "functional literacy" since functional literacy emphasized obtaining primary economic and social skills and ignored the requirement of developing attitudes and values (Ordonez et al., 1998, p. 51).

The investigation and survey for this study showed that the main reason why the farmers were in poverty, and were conservative-minded as well as having high levels of technological illiteracy is because the rural or mountainous areas remained sealed from the outside, kept isolated, and impenetrable for a long time. This isolation referred to the farmers geographically, intellectually, and with infrastructure. For instance, fewer roads are available for access to public service such as health, education, and other extension services (Ordonez et al., 1998, p. 41). This spatial isolation results in the intellectual isolation in terms of a low education level, technological illiteracy, as well as conservative-mindedness and the ignoring of technological consciousness. In earlier interactions with the farmers, it was a common case for farmers to stick to the traditional methods and refuse to change to scientific farming practices. For example, at the beginning of the Taihang Mountain Development project, when the professors were to prune the fruit trees, the farmers said, "in that case where will the fruits grow?" The well-bred genetic maize given out to the farmers for planting in their fields were instead privately fed to chickens, etc.

With farmers showing resistance and unresponsive behavior to change resulting mainly from ignorance, professors and staff of AUH had to conduct experiments on selected backbone farmers' land and on abandoned fruit tree areas and farms. The farmers were not sincerely convinced until they saw the harvest on the pruned persimmon trees and the well-bred maize production was doubled.

Given this experience and with the knowledge of the professors and staff and witnessing the developmental fact. the farmers understood the power of technology and the reality of "eliminating ignorance" ahead of "transforming the mountain." They were convinced that only if they had a strong technological consciousness, could they truly advocate scientific ways. From such consideration only they could build the ideological foundation for technological literacy.

The connection between "transforming the mountain" and "eliminating ignorance" was firmly set, and the technological consciousness of the farmers was raised; the farmers sought more willingly than before for more technical help from professors, staff, and students of AUH who were highly respected now and regarded as "Plutus." Actually, "Plutus" is knowledge, technology, and is their own empowerment through knowledge and technology

An analysis of statistical data revealed that each year the university sends out technological session teams consisting of more than 100 experts/professors and over 400 students of various majors. They immerse themselves into the rural communities and through practical experience, technical guidance, and service carry out technical consultation and training activities. Every year AUH holds over 400 training seminars attended by more than 50,000 trainees. It also distributes more than 10,000 copies of practical technology materials and technical reference books. This is so that the low-level elementary education populace or grassroots farmers could acquire the necessary technical knowledge and skill and rapidly master one applied technique. Besides vocational training, adult training, technical training, and green certificate, some other intensive formal training was also carried out and the diploma or green certificate awarded.

Poverty in Rural, Remote, and Indigenous Communities of NT

Apart from a few cities and towns, a broad area of NT is referred to as rural, remote, and Indigenous regions. People living in these areas "tend to have lower levels of household income and higher levels of unemployment. In most county regions, unemployment levels are higher than in the capital cities" (Wyn et al., 2002). And also "The evidence pointed to the generally lower incomes of those living in these regions; reduced access to services such as health, education, and transport, and declining employment opportunities"(Committee Hansard, August 4, 2003, p. 1193 [QCOSS]). Furthermore:

> NTCOSS (NT Council of Social Service) reflected on the specific characteristics of the Northern Territory which impact on poverty in the Territory, including remoteness, a large Indigenous population, the problem of distance with a small population spread over a large geographical area and high population mobility. NTCOSS noted that these factors "pose challenges in providing adequate physical and social infrastructure as well as cost for people in the NT." (Committee Hansard, 2003, p. 1081 [NTCOSS])

There is a voice from the rural communities: we are in poverty, because we are in drought. Our kids are educated out of our communities, and cannot return; we cannot have doctors and banks; no train or other public transport (Committee Hansard, 2003, p. 1066 [Han. Jonathan Ford MLC]).

Rural Indigenous communities have more problems. For example, people's life expectancy is 20 years less than that in non-Indigenous regions and has twice the mortality rate of non-Indigenous regions (Submission 148, p. 29 [Catholic Welfare Australia]).

Capacity building, education, and training are considered as the means of addressing poverty within the communities. It was noted that is it was important to "build community networks so as to improve social capital, and to strengthen the capacity of communities to deal with the consequences of hardship" (Submission 148, p. 30 [Catholic Welfare Australia]).

Commonwealth and NT governments and to some extent the NGO have implemented various community development programs with strong regional dimensions in such areas as health, education, housing, and transport, as well as other community services, such as banking.

One of the greatest concerns in the NT's poverty situation is the considerable group of Indigenous population, which occupies 28.5 percent of the total population in the NT (Chapter 5). On every social and economic indicator, Indigenous people are at the lower end of the scale. They have the highest birth and infant mortality rates, the highest death rate, the highest imprisonment rate, the worst health and housing, and the lowest educational, occupational, economic, and social status of any identifiable section of the Australian population (http://www.sa.alp.org.au/policy/platform/indig.html, accessed on June 16, 2011).

Lack of human resources exists in most Indigenous communities. It is apparent that education and training inside communities can make a positive contribution in terms of human resources development. Also, another solution is to "create a home environment that attracts the more talented and educated people back home" (http://www.abc.net.au/darwin/stories/s980849.htm, accessed on April 16, 2011).

Increasing the capacity of Indigenous communities to determine their own pattern of development relies heavily on education and training, this educational intervention has been put into practice. For instance, 17 Indigenous and 15 non-Indigenous essential services officers were presented graduation certificates for the Certificate Level II in Electro technology Remote Area Essential Services Operations (ESOs) by the Member for Karama Delia Lawrie, recently.

In two senses these guys are really in the front line for their communities. First, because they are the ones on call to deal with essential service emergencies that might arise, whether it involves water or power supplies. Second, because they are graduating into highly skilled, real jobs so vital to boosting employment status on those communities. In order to increase Indigenous participation in essential service delivery, Power and Water Corporation are supporting a training program with a local Group Training Company, Group Training NT. These trainees will become the skilled, locally based Indigenous people who will be the ESOs of the future, not only in the major communities but potentially in providing services for the 600 remote outstations. While the outcomes from this ESOs course will ensure that remote communities enjoy fewer service interruptions with faster response times to rectify faults and a better understanding of conserving resources, the course represents a successful partnership between Indigenous communities, private training providers, and government. It is a major achievement within the broader objective of Indigenous employment and economic development and shows the potential of local Indigenous people to become involved in real jobs, act as role models, and to better participate in the economic development of their communities (http://www.nt.gov.au/ocm/media_releases/2004/20040325_essential_services.shtml, accessed on April 16, 2011).

Apart from capacity building, education, and training, another and also important measure to deal with Indigenous population and for poverty alleviation in Indigenous communities is to recognize or acknowledge the potential of Indigenous knowledge, culture, and skills. This has been accepted by many governments, university staff, and community workers.

Summary

In the areas, such as rural or mountainous regions of Hebei, it is apparent that poverty appears as less income and lack of a substantial life or access to materials goods. This poverty is only considered as income poverty. As mentioned earlier, there are still other dimensions of poverty, for example, capacity poverty (lack of basic capacities).

And in rural, remote, or Indigenous communities of NT, the study shows that the poverty is mainly capacity poverty.

A sustained poverty alleviation program targeting the rural poor should focus on: (a) household food security; (b) nutritional security; (c) livelihood security; (d) economic security; (e) environmental security including health, sanitation and environmental management; (f) ecological security including bio-diversity consecution; (g) human resources development through skills; and (h) social and gender equity. The universities, as higher education and public service institutions, have a responsibility of subscribing to and implementing the "Declaration on Food and Nutritional Security" adopted at the World Food Summit (1996).

Education and Poverty Alleviation

The right to education is recognized as a fundamental human right. Better education is also an important development goal. It clear that good education is also a means to achieve other development goals related to poverty reduction: that the linkages between education and poverty alleviation are powerful and much stronger than is generally understood.

Education in its broadest sense is a precondition for poverty alleviation and sustainable development. Education contributes to economic development and poverty alleviation through such things as improved learning in childhood and adolescence, positive demographic changes, increased productivity, and increased the opportunities on income-generation. It is also noted that investment in education and training must be accompanied by investment in other factors, such as health, water and sanitation, and agricultural improvement to achieve development outcomes and poverty reduction.

The goal of poverty alleviation through improved education can only be effectively addressed by means of a decentralized approach that ensures community empowerment, local institutional development, and promotion of good governance. Well-functioning educational systems require incentives that attract qualified personnel, adequate infrastructure and information systems, and capacity for planning and effective leadership. Access to basic learning needs at different levels, is also a precondition for the functioning of health systems. Together, these are essential conditions for the sustainability of educational systems at country level. And it may be argued that sustainable educational systems are an important condition for reaching the objective of sustainable development overall.

It is clear that education as a vehicle can:

> [f]urther enhance the productivity and efficiency of farmers, facilitates the development of the non-farm agricultural sector, and enable rural households to respond to rapid changes in technical and economic conditions and take advantage of new opportunities. (World Bank Report, 2001, p. 2)

And also it is generally agreed that "education constituted as at present is not specifically designed to alleviate poverty. Objectives relevant to poverty alleviation are not addressed" (Ordonez et al., 1998, p. 44). But education and lifelong learning, aimed at raising the awareness of the poor on aspects of the life, can be seen as the part of the foundation for poverty alleviation and improvement of quality of life (Ordonez et al., 1998, p. 12). Therefore, it is well acknowledged that education, as a useful process of working on human being in order to bring about desirable changes in the way that we think, feel and act, can empower the poor by attacking ignorance through providing information and knowledge, building skills necessary to change their life, changing attitudes and values, and meeting the learning needs of the poor (Ordonez et al., 1998, pp. 4–16).

Development and Rural Development

Development and Rural Development: Prospective Future for the People in Rural, Remote, and Indigenous Areas

Development and rural development can be realized through economic growth, which can effectively reduce poverty and improve people's living standard. But economic growth is not only a reflection of development and cannot be realized without the consideration of human potential development in terms of its material, spiritual, individual, and social dimensions. Attention has been drawn to the factors, which affect human development. "Human development is a process of enlarging people's choices. Three essential areas are for people to lead a long and healthy life, to acquire knowledge, and to have access to resources needed for decent standard of living" (UNDP, 1991).

Atchoarena and Gasperini highlight rural development in their recent published book: *Education for Rural Development: Towards New Policy Responses* as:

> Rural development: encompasses agriculture, education, infrastructure, and health, capacity-building for other than on-farm employment, rural institutions and the needs of vulnerable groups. Rural development aims at improving rural people's livelihoods in an equitable and sustainable manner, both socially and environmentally, through better access to assets (natural, physical, human, technological, and social capital), and services, and control over productive capital (in its financial or economic and political forms), that enable them to improve their livelihoods on a sustainable and equitable basis. (Atchoarena and Gasperini, 2003, p. 21)

Education is a social practice for human resources development and national capacity building. Also, education, as a medium for empowering human being, providing knowledge and information, increasing farm productivity and farmers income, reducing fertility rates, enhancing health and nutritional factors of human beings, and other quality of life indicators, as well as changing attitudes and values, can make contributions for rural development. To some extent, education can play centre stage of development, and from this point of view, it is as an ethical imperative, that "education is an intrinsic good to be cherished for its own sake and a basic human right" (Ordonez et al., 1998, p. 5).

It is a common understanding that communities cannot be developed without an educated population. And it is clear that educated farmers are willing to accept new knowledge, technology and information, and actively participate in the programs for promotion of social, economic, and cultural aspects of development. Also, without basic knowledge such as fundamental literacy, numeracy, problem solving and productivity skills, rural excess labor find it difficult to obtain a good wage work outside the farm, whether in rural or urban areas.

Sometimes, education/ training and development, especially economic development, as well as social improvement in rural areas seem to be out of phase.

> Businesses, large or small, are unlikely to choose to invest in rural areas if skilled or trainable human resources are unavailable. Similarly, a community cannot retain educated people without an attractive economic environment. Many poor rural areas, mostly but not only in developing countries, are trapped in this situation. (Atchoarena and Gasperini, 2003, pp. 28–29)

Education and training are powerful vehicles for human capacity development and for rural or community development. Furthermore, there are many reasons to believe that education contributes positively to improve people's life styles and living standard. Education contributes also towards increasing a farmer's technological consciousness and awareness of agricultural knowledge, and information to impart positively on their agricultural productivity and community development as well.

Knowledge, Technology, and Information Transfer from University to Rural Communities for Development

Universities as higher educational institutions have many advantages in terms of human resources of staff and students, advanced knowledge, information and technologies, and mechanisms to deliver them into different bodies. Universities also "[have] a responsibility to provide teaching and learning opportunities for those who seek careers in the management of the rural development process or who will, at various levels, implement … rural development activities and processes" (Atchoarena and Gasperini, 2003, p. 311).

What can the universities bring to knowledge, transformation, and education for rural development?

> It can educate the professional and technical personnel needed to promote sustainable agriculture and take leadership in implementing the process of rural development. It can bring critical agricultural messages to the education system at primary, secondary and adult levels. It can tap into the desire of millions for lifelong learning. It can be the voice of reason and factual information in emotional debates about real or apparent food quality and food safety issues. It can equip teachers of the education system with the knowledge and skills required to bring the agriculture message to that system's enrollees. It can be an invaluable resource for policy-makers. (Atchoarena and Gasperini, 2003, p. 332)

It is clear that the practices of CDU and AUH, discussed in previous chapters, have shown this transformation successfully. The followings are summaries of their stories.

In AUH, the rural development services have been in area, such as to guiding the counties, townships, and villages to formulate their social and economic development programs; implementing the joint development project of production, teaching, and research; delivering various forms of education and training; and facilitating village leaders and farmers to acquire new ideas and concepts of development. Due to its significant contribution to the development of Taihang Mountain areas, the Chinese government has called its experiences "the Taihang Mountain Road," which should be followed by others (Chapter 4).

The key features of AUH involvement in rural development programs are: integration of teaching/training, research, and production; using multidisciplinary techniques and knowledge to develop a comprehensive regional development plan; strong leadership, technical and backup services; influencing farmers through model households; and student volunteers' intervention (Chapter 4).

In CDU, the knowledge transformation and rural development programs focused on empowering the local people with knowledge, technologies, and information in rural, remote, and Indigenous communities. The intervention includes: education/ training, health, Indigenous education, research, and consultancy, as well as agricultural/ horticultural extension (Chapter 6).

More Concerns on Rural Development

There are many factors that affect rural development. Education alone cannot be responsible for rural development. There are still some factors, like economy, culture, tradition, social stability and security, resources, natural conditions, and so on, that effect the development of a rural community. Education is one of the important issues to empower the rural people for development.

> Empowerment is the process of increasing the capacity of individuals or groups to make choices and to transform those choices into desired actions and outcomes. Central to this process are actions which both build individual and collective assets, and improve the efficiency and fairness of the organizational and institutional context which govern the use of these assets. (http://lnweb18.worldbank.org/ESSD/sdvext.nsf/68ByDocName/WhatIsEmpowerment, accessed on April 16, 2011)

Empowerment is an important issue. It can aid self-determination. Empowerment can change the environment in which poor people live, helping them build and capitalize on their own attributes. Educational activities which empower poor people are expected to increase development opportunities, enhance development outcomes, and improve people's quality of life.

Therefore, empowerment of rural poor can become a main issue for rural development, and education is a key measure to empower the poor. In terms of empowering farmers and improving their literacy level through education and training, 75 percent (12 out of 16) interview results show that direct on-site teaching and training is a best way. This feeling is best summarized by the following extract:

> Farm forums or farm discussions experts/panelists go to the sites and have dialogue with farmers about their farm problems. I have been actively involved with these for years working in the applied communication office where we have the so called agri-caravan where for a quarter of a year we moved from one place to another to hold a fiesta-like activity showcasing the best produce by farmers, a contest linked with the technology hands-on demo where farmers who could perform the techniques very well have the chance to win the prize at stake. The prize maybe in the form of fertilizers, farm chemicals or live animals such as piglets or goats depends on what our donor provides. Then a panelist of local leader from the mayor or governor's office together with technical experts will conduct a dialogue among

local people raising technical issues and other related ones. (SS 7, 2002) (I have decided to include the exact references to the survey data so that readers can get a feel for the responses directly from the survey instrument that I used. I will code the responses SS so that readers can cross check the survey responses to the questions and should they wish to, develop follow-up surveys to extend the research.)

The results of interviews have shown that the demonstration is very important for the farmers in China (77.8 percent, 7 out of 9), since:

Most effective way to extend new varieties and new skills for farming is to show the results of new varieties by carrying out experiment in rural areas. As farmers in China cultivate their land individually, they only use new varieties and new skills for farming if they see personally the practical results (SS 2, 2002).

Of course, "extension activities, but in a new approach" should also a concern for rural economic development (SS 12, 2002 and SS 7, 2002). Furthermore, other activities related to rural development should:

[a]ctually relate to the needs of the people living in these communities (now and future); to improve lives of rural people, economic, health, social and cultural; to involve the people in planning, and implementation; to empower people; economically, politically, socially. (SS 1, 2002)

Other concerns from surveys in this regard could be summarized as: in terms of educational issues, community members' self-consciousness is critical, and a way for them to "understand their real situation and take informed action to make changes" (SS 8, 2003). The community members may request appropriation workshops, training courses, such as "building construction training" and demonstration activities (SS 8, 2003, SS 10, 2001, and SS 11, 2002). And also: "Bringing in rural community leaders, with whom they can relate, is a way of exposing new ideas to a given community, without them feeling that the educational institution is imposing new ideas on them without regard to their actual needs" (SS 3, 2002).

In terms of rural transformation activities, most (81.3 percent, 13 out of 16) agree that it is a long-term changing process and education plays a vital role. The surveys show that those should be something to "help people develop human capacity, hence improve community capacity to achieve social and economical gains" (SS 5, 2002).

Some other surveys highlight that:

> Changes are not going to happen immediately. There is a lag period. It takes some time. It depends on the total attitude of the community. Some resist change. Some welcome change. But normally there are groups of people (often they are key persons) who will have great influence on the changes that will happen. If the attitude is lukewarm, the development worker should pay more attention to this key people to convince them by letting them understand the project. If they resist—take your time but persist—they are observing you carefully. If their attitude is positive—make full use of them by making them fully participate—making them key implementers of the changes and give credit to them. (SS 7, 2002)

and

> This can have positive as well as negative effects. I am worried about negative rural transformation, where rural communities are transformed into poorer and further disadvantaged communities, where social and cultural values and language are lost and where family structures break down. Positive rural transformation is the opposite. It is about building the quality of life in rural communities without losing the good qualities of rural communities. Rural transformation involves changes in attitudes, and practices. Education plays a vital role. (SS 1, 2002)

Many surveys address the important issue for rural transformation activities to be successful that they must be sustained when the project implementers leave the project sites. There should be "a program in place which gives great emphasis to follow-up communication and feedback. 'Show-and tell' and 'hand-on' are best" (SS 6, 2003), "training local people as team leaders or project coordinators to carry on some of the activities" (SS 5, 2002), and "should be a policy of government rather than an individual action" (SS 2, 2002).

Further extracts show that some policies experiences should be used, for instance:

> Before leaving the sites, capability training is a must, which means a set of rural community leaders must oversee the project to sustain its success level. I have been involved in one assessment of the "Save the Children Foundation" project site. SC Foundation after implementing the project has properly trained key leaders to continue managing the cooperative that has been running successfully for years under their guidance.

In my own experience I think regular visits to call on the key leaders and some home families will make them feel they are remembered. This can be coupled by attending their activities once in a while. I have been invited to many social functions in communities where I worked before. I always find time to attend this. They also appreciate that because they feel they have acquired sustainable links to my organization.

But in my own theory what is more important is to link rural trans-formation to the economic activities of the people in the community. If these changes will provide them opportunities to continually better their lives—the changes that have been initiated will take care of themselves and transformation process will continue. In principle that means rural industri-alization is the best sustainable approach—by leaving a significant industry (in the form of enterprises run by the people) to provide jobs and improve the economic life of the residents. Economic development will enhance social transformation and will better the health system as well the environ-ment of the communities. (SS 7, 2002)

Concerning what kinds of technical support, knowledge, skills and other approaches need to be used to increase farmers' (or rural workers') scientific knowledge and capabilities, the answers are quite different due to their personal feelings, experiences, and knowledge bases. For exam-ple, some survey respondents believe that a series of methods could be used, such as:

(a)Actual demonstration followed by hands-on training; (b) Video tapes of techniques that farmer can see over and over again; (c) Farmer scientist approach—a farmer who has original ideas has been helped to package his technology with the help of the technical personnel. (The farmer teaches the technical personnel he in return understands the technology and packaged it scientifically for further dissemination); and (d) Demonstration sites— where proven techniques were used in the farmer fields and he is guided how to use the technology. Later he will become the core source of tech-niques that will radiate in the whole community. (SS 7, 2002)

Onsite training, showing, and demonstration should be a more useful approach, which shows farmers how to do it and explaining the impor-tance and economic benefits of such practice. Because, "It is just like selling a product. You sell the product not because you want their money but you try to provide the benefits the product could offer to them. In sharing technology, it is like selling—you should make the people feel that you only want to help them make their lives better with the technology" (SS 7, 2002); other argues that easy understanding manuals or reading materials with simple language, using local experts or local farmers to help explain things, formal training courses, demonstration and exhibits, and agricultural shows should be useful. (SS 6, 2003, SS 16, 2003, and SS 1, 2002)

The Roles of University for Rural Development

Why Universities

Universities are higher educational institutions. The World Conference on higher education in the 21st century: Vision and Action, was held in UNESCO Headquarters in Paris, from October 5 to 9, 1998. The conference report pointed out that: "the core missions of higher education—to educate, to train, to undertake research, and to provide services to the community—must be preserved, reinforced, and further expanded." The conference also stressed that "higher education institutions must seek to educate qualified graduates who are responsible citizens and to provide opportunities for higher learning throughout life" (http://portal.unesco.org/education/en/ev.php-URL_ID=19294&URL_DO=DO_TOPIC&URL_SECTION=201.html, accessed on April 16, 2011).

The purposes of higher education in Australia "The Government regards higher education as contributing to the attainment of individual freedom, the advancement of knowledge, and social and economic progress. The main purposes of Australian higher education are to:

1. Inspire and enable individuals to develop their capabilities to the highest potential throughout their lives (for personal growth and fulfillment, for effective participation in the workforce and for constructive contributions to society);
2. Advance knowledge and understanding;
3. Aid the application of knowledge and understanding to the benefit of the economy and the society;
4. Enable individuals to adapt and learn, consistent with the needs of an adaptable knowledge-based economy at local, regional and national levels; and

5. Enable individuals to contribute to a democratic, civilized society and promote the tolerance and debate that underpins it." (Karmel, 2001)

The significant task of higher education in China is to foster highly skilled personnel with the spirit of creativeness and the ability of practice, to develop science, technology and culture, and to promote modernization drive (UNESCO Principal Office in Asian and the Pacific [UNESCO PROAP], 1998, p. 21). It includes education for academic qualifications and education for nonacademic qualifications, and also takes the forms of full-time schooling and non-full-time schooling (The Laws on Education of the People's Republic of China, 1999, p. 92).

What Is the Higher Education Function?

Generally speaking, three main functions need to be fulfilled by the university: teaching/training, research, and extension/consultancy. However, these functions can be expanded to include specifically social and economic development. The Asia and Pacific Regional Conference on Higher education National Strategies and Regional Co-operation for the 21st century held in Tokyo, Japan, (July 8–10, 1997) made the Declaration about Higher Education in Asia and the Pacific, that is:

> [W]e reaffirm that the aims of higher education can be summarized as followings: to educate responsible and committed citizens, to provide highly trained professionals to meet the needs of industry, government and the professions; to provide expertise to assist in economic and social development, and in scientific and technological research; to help conserve and disseminate national and regional cultures, drawing on the contributions from each generation; to help protect values by addressing moral and ethical issues; and to provide critical and detached perspectives to assist in the discussion of strategic options and to contribute to humanistic renewal.

The Roles of University to Serve the Local Economy, Especially Rural, Remote, and Indigenous Communities, and Transformation Models

In terms of the ways in which the universities can serve the rural communities, some arguments emerged from surveys and interviews carried out during this book's development. For example, one survey mentioned that:

> There are three ways for universities and higher education institutions to serve rural development. The first is to develop new varieties and new skills for farmer according to the practical needs. It has proved that technology constitutes the key to development of modern agriculture. And new varieties and new skills for farming cannot be born among farmers-universities and research institutions of agriculture should undertake this task. The second is that universities and higher education institutions should help with setting up network of higher and secondary educational and technical institutions to extend new varieties and new skills for farming. The third is to send professors in agriculture to rural areas to help solve the problems farmers encountered. Though it is not possible for professors to spend much time working in rural area, but being present in fields with farmers will have exemplary role for other farmers. (SS 2, 2002)

The others mention some ideas following the role of university, for instance, "The University's role is to bring new ideas, better practices, and recent discoveries to rural communities in order to improve the level of production and rural living standard" (SS 6, 2003). Another concern is that "extension service" should be the best way to serve the rural communities, which includes extending technology or promoting technological literacy among the rural folks. Furthermore, capability building by making farmers literate enough to make choices and wise decisions also needs to be concern, which could provide farmers survival skills (emergency and health knowledge e.g. first aid), functional skills (e.g., some skills to get a job—technical skills), etc. (SS 7 2002 and SS 13, 2002). Some people believe that collaborative work should be used to meet rural people's educational, research, and consulting needs. Rural development should be long term and should have a "better match between what the educational institute provides and the real needs of the rural community" (SS 3, 2002).

Others emphasize that universities have a role in serving their community, which is a responsibility for the universities located in regional or rural areas to all people in their location. Universities located in a rural community should be a main vehicle to help support the community and local development and community survival, which is an issue for rural transformation (SS 15, 2003) and a "concern where communities are getting smaller and urban centers" (SS 1, 2002). A lot of surveys were concerned that the universities' rural development or service programs should be "regular, on-going, on-site, and face-to-face contact with all providers and students"; "community based workshop" (SS 5, 2002) should be used to "providing specific advice and assistance in the development" (SS 8, 2003); "introducing new techniques" (SS 4, 2003), etc.

In terms of the strategies in which the universities can use to translate technical knowledge from the universities to rural communities face-to-face showing and demonstration is critical since "I think that when people see success they are more inclined to want to try something rather than just hearing the story of something ... This is particularly the case in poor areas" (SS 1, 2002). Living in a rural community and participating in their activities have proved to be effective strategies (SS 7, 2002).

> This was a very risky experience especially when I once lived in an area frequent by rebels because I am being watched and also my activities. But my dedication paid off. I was not only able to change the old beliefs of people in their farming practices but I also gained their sympathy and they empowered themselves. I have a face-to-face talk with some sympathizers of the rebels and the rebels themselves. I don't know how I win over them but I have prevailed. (I just told them that we maybe similar aim to liberate the poor the only difference is the way to achieve it–I shiver with fear when I remember this encounter).
>
> When I went home this January—I made a visit to that place and saw some changes—I don't claim such changes for my effort but what I am happy about is that I have been part of the early development of that community (SS 7, 2002).

"Any activities which will help the farmer improve production levels and therefore family income, as well as a higher standard of living for the rural community" (SS 6, 2003) could be considered as useful for universities. Most of surveys show that almost all of the following are useful: training workshops, reading materials, demonstration and exhibition, on-the-job training, project-based learning, apprenticeship, study tour, interest group, hobby group and radio, TV, posters, manuals, and other media.

Table 9.1 shows that, according to surveys, all activities named here are helpful; four items are very helpful, that is, training workshop (62.5 percent), demonstration (75 percent), on-the-job training (75 percent), and project-base learning (56.3 percent).

Concerning the strategy to be ranked as the number 1 in importance in getting university knowledge and research into rural communities,

Table 9.1
Statistical Data on Helpful Activities for Universities for Rural Development

	Not Helpful	Less Helpful	Helpful	More Helpful	Very Helpful
1. Training workshop		2(12.5%)	3(18.7%)		10(62.5%)
2. Reading materials		2(12.5%)	4(25%)	1(6.3%)	6(37.5%)
3. Demonstration and exhibition			2(12.5%)	2(12.5%)	12(75%)
4. On-the-job training			1(6.3%)	2(12.5%)	12(75%)
5. Project-based learning		1(6.3%)	3(18.7%)	2(12.5%)	9(56.3%)
6. Apprenticeship		1(6.3%)	7(43.7%)	4(25%)	3(18.7%)
7. Study tour		3(18.7%)	7(43.7%)	5(31.3%)	
8. Interest group		3(18.7%)	5(31.3%)	5(31.3%)	2(12.5%)
9. Hobby group		8(50%)	3(18.7%)	2(12.5%)	2(12.5%)
10. Radio, TV, posters, manuals and other media		2(12.5%)	11(68.7%)	2(12.5%)	

Source: Data from surveys.

most of the Chinese surveys shows that project based learning, demonstration, and extension are very important in this regard, and:

> Generally speaking, agriculture universities in every country know that their work should serve agricultural development in each country. However, they are thinking of more in human resource development, less in extending the findings to rural areas. There is a traditional concept that university people should do research, extending should be done by others. Therefore, that the extending of findings should constitute part of work for university has not been solved in many developing countries. This is in concept. More importantly, it is more difficult to find practical ways that are acceptable to all agriculture universities as to effectively extend the findings. The most important strategy, therefore, is to have an agriculture university in a country to do pioneering job in this respect. Their successful story in extending the finding of university would encourage other universities to follow its example. (SS 2, 2002)

In terms of essential factors for successful translation of university findings into usable rural information, my interviews with key informants from China show that most respondents in order to make better use of human resources, and better services for rural communities, agricultural university should be set up in rural areas, at least some areas closely linked with rural communities, or set up some campuses in rural areas. Other arguments highlight some areas, such as, simple extension programs, face-to-face talking by university professors and staff, as well as training workshops and other courses.

Responses from key people interviewed in Australia are quite interesting; many of them think that the information or findings should be closely related to the need of local area and must be relevant to the farmers (SS 2, 2002; SS 6, 2003), there should be "provision of information on the application of the research in a manner that nonscientific people can understand" (SS 4, 2003), "making information accessible to people with low levels of formal education" (SS 8, 2002), and "written in language that is appropriate to the implementaters of change. Much university research is written for a university audience and follows an academic format that isolates rural people. Universities are still caught up on this elitist model of higher knowledge that is power" (SS 1, 2002). Other concentrated opinion emphasize: make sure farmers' perspectives are included and research actually closely relates to the need of rural people, work with or involve rural people, work as equal partner or have joint ownership in a research project, research should be allowed with practical application in real project, and so on.

From previous discussions and information from previous chapters, apart from teaching/training and research, the university reveals that a key function, which is a significant factor for the university to transform knowledge to rural communities for their development is that of extension training. Evidence that services in human resources development and human capacity building in co-operation with different government departments and agencies for local economy and rural development is an emerging interest.

Three important issues relate to the successful transformation of knowledge base from universities to rural communities. That is: universities themselves need to clearly identify that they wish to make a contribution for rural development; governments in different levels need to commit to support universities' rural service programs; and communities need to participate positively in universities' rural development programs.

Several models of university transformation practices emerge. From the analysis, several models can be developed that identify the aspects necessary for successful university intervention in rural development.

Model one: The important model for the roles of university for rural development is to establish a lifelong partnership with rural communities. Individuals then have an opportunity for lifelong learning. These options should be flexible, including formal and informal training. Training courses in rural areas could be many one-day courses. Universities should become "skills banks" for each individual in which the student banks their skills and knowledge over a long period of time allowing universities to develop flexible qualification certification requirements. Universities need to establish a long-term learning community in rural areas. "Clearly there is much to be gained from such a partnership, including a strong focus of university resources and expertise on social, cultural, and economic issues regarded as important by the Territory Government, and the opportunity to strengthen the university in areas critical to the future of the Territory". Professor Ron Mckay, former Vice Chancellor of NTU (a former name of CDU), said (On Campus, March 2002, p. 2, Vol. 4, No. 5, March 2002), "Of course, there is nothing particularly new about the idea of working together for mutually beneficial outcomes. There are many examples of successful joint projects involving the university, government departments, and others across a range of areas in education, training, and research." This model requires continual reinforcement to be successful.

Model two: To combine theory with practice to make benefits for the students and local communities members. For instance, the CDU practice firm, Crocodylus World, sponsored by Crocodylus Park, a private company, is a fine example. Professor Ron McKay, former Vice Chancellor of NTU (a former name of CDU) addressing on the 2002 Top End International Trade Fair for practice firms or virtual enterprises, had the following comments (On Campus, July 4, 2002, Vol. 4, No. 12). The idea is that: "practice firms have simulated workplaces in which students learn about business by doing it in a safe environment, allowing skills and abilities to be tested and developed." Some other cases from CDU and AUH in this context have also proved that it is possible and desirable to combine theory with practice to be an efficient and effective way not only for university but also to make a benefit for the rural community.

Model three: Government commitment, financially and administratively, has proven to be a very important ingredient for the university to be successfully involved in rural and regional development programs. For instance, the Northern Territory Government announced on April 30, 2004 that it would put an AU$2.5 million commitment to CDU's Alice Springs Campus, which will enable the immediate establishment of a higher education center in this campus and make higher education in Alice Springs a reality. As a result of this initiative, better facilities and greater higher education opportunities for Central Australians will be possible. Besides, a further $500,000 will also go towards integrating IT systems, stationery, the website, and other system changes.

"One of the main aims with the creation of CDU was to ensure that it was a University for all Territorians, no matter where they lived. This investment in Central Australia will help to achieve that vision. Importantly, it means more Central Australian students will be able to pursue their studies locally, rather than having to move to Darwin or interstate and staying there."

"It demonstrates a commitment to build local capacity in higher education and research, particularly in desert knowledge, to underpin the region's social and economic development."

Importantly, a wider range of lecturers and course programs will be offered through an integrated e-learning component allowing students to simultaneously participate in classes delivered from the Casuarina (Darwin) campus, 1,500 km to the north.

As well as enhancing the delivery of programs to "on campus" students, the Centre will provide a resource base from which educators can deliver programs to regional communities (http://www.cdu. edu.au/newsroom/stories/2004/april/healice/index.html, accessed on April 30, 2011).

Model four: Has its focus on community participation. This is another key issue to ensure that rural development programs reach their expected goals. For instance, during AUH implementation of the project of Revitalizing Villages through Science and Education, a professor came to a village and lived with the farmers to deliver the innovative techniques. The first time when this professor conducted the technical training class, no one participated even though the village leaders announced several times. After further motivation by the leaders from door-to-door encouragement and promising to pay those who attend the training course, finally, eight villagers came in. After training, these eight villagers realized the importance of technology in their agricultural production. Soon they began to encourage others to attend the training class voluntarily. Gradually, more and more farmers came to classes, even those farmers from the nearby villages participated in this activity (INRULED, 2000).

The Future Perspectives to Enlarge the University's Rural Development Programs and Make the Benefits for Rural People

Besides the traditional measures for universities to transform their knowledge bases to rural, remote and Indigenous communities, modern information technology and digital media such as e-education/training, e-commerce, and tele-communication increase the possibilities for universities to enlarge their rural development service and programs. These innovations will provide considerable promise to enhance teaching and learning for both on campus and distance education and training. The approach will help to establish networks of institutions and scholars, to serve larger groups and to facilitate communications among researchers and teachers. Of course, at the same time, harnessing this technology will require considerable investment in both hardware and staff development (Declaration about Higher Education in Asia and the Pacific, Higher education National Strategies and Regional Co-operation for the 21st century, July 8–10, 1997).

Both the case study universities AUH and CDU have further considered and used this digital opportunity to serve the rural population. Through innovation and partnership building with government agents, relevant organizations, and communities, the universities have been able to transform their research knowledge for rural development.

What are Main Similarities and Differences of the Activities Carried out by CDU and AUH in Rural Development

In the view of traditional activities to serve rural communities for rural development, the two universities, AUH and CDU, have done almost the same interventions, such as teaching, training, research, consultancy, and extension, as discussed in previous chapters. Comparatively speaking, some weakness or strong point appear in different areas for each university's rural service programs, only because the priorities focused on by two educational institutions and conditions for both countries are different.

Other similarities, for example, establishing a partnership with government as the way to help facilitate universities in supporting the development of the rural communities; community participation ensures the program obtained the expected outcomes.

The different approaches adopted by the two universities are a product of different environment, conditions, and different service bodies in the rural areas. For example, as an agricultural university, AUH has concentrated more in agricultural technique extension. At the same time, as a developing country with a majority of its population living in rural areas, urgent educational needs of rural Hebei required AUH to focus its activities on agricultural and related areas; income generation for the rural population; technology illiteracy alleviation; promotion of rural people's life; and so on. All of these approaches are much more urgent and important issues to be considered by AUH. Furthermore, the traditional media is still in a main channel for AUH's rural development programs, but its digital services and information technology are in an initial stage to be used for rural development programs.

CDU is different. As the biggest Indigenous population in Australia lives in the NT, CDU has made considerable contribution to Indigenous education and training as well as Indigenous Community development.

Another difference is that most programs of CDU delivered in rural areas are aimed at human capacity building in education, health, environment, financial, etc The third difference is information technology and digital media. This has been broadly used by CDU for its rural, remote, and Indigenous education, training, extension, and so on.

Summary

Universities can provide information, knowledge bases, facilities, and talented staff; governments have responsibilities for financial, administrative supports and relevant strategies, and policy-making; communities need to have a positive attitude to ensure the motivation of their members and their participation, and the provision of local services. Modern digital communication and information technology are important contributors for knowledge transformation and technology dissemination. All those efforts aim at the establishment of a learning society to empower population in rural, remote, and Indigenous communities for development.

There appear to be five important aspects for knowledge transforming activities of universities to be successful.

First, the university itself. There is an urgent demand for the universities, especially agricultural universities in developing countries to concentrate their activities for rural communities in terms of technological extension, training, and local people's capacity building. This is necessary in order to improve their communities' living standard and well-being, and to benefit the university itself.

Thus, these kinds of programs enable the university's teaching, research, and other academic activities to be used for the mutual benefit of the universities and local communities. Besides teaching, learning, and research, university's staff and students should be involved in extension programs, taking the university knowledge base to the community.

Second, rural community participation is another important aspect of rural development. The rural development programs carried out by the universities or other agencies must be attractive to the people both psychologically and economically to make the programs more acceptable for the local people in the community.

Third, government commitment and actual involvement in the programs enable some of the weakness of universities to be overcome, for example, government funding is an essential part of success in community development programs.

Fourth, modern information technology and digital media give universities a great potential and prospective opportunities to be involved in rural development program effectively and efficiently.

Finally, the program implemented in rural community itself, should be more sustainable to make sure that it could be still active after the program implementer has left the program site.

There is no doubt that the universities in general have a mission, function, or role to serve rural communities. Do professors, staff, and students themselves, have such a function?

Program Needs

Some programs (research, training, and extension program) need to be carried out in rural areas, especially agricultural university's programs; therefore, serving a rural community is an integral part of the program objectives.

Administration Needs

Exemption from some restrictions for promotion, for example, in a Chinese university, everyone needs to have a English exam before promotion can be gained, it is compulsory policy, but in some cases, if the staff serve in a rural area for two years, this exam is waived.

Student Needs

University students in some areas, especially agricultural university students, need to have a period time, at least six months, for practice to meet the degree requirement.

University Needs

Universities also need to have their extension bases developed so that new varieties could be tested; new research areas could be found; and also new research could be carried out.

This research has shown that the university's extension advice and service provided meets some of the needs of farmers and rural communities. This particularly applied to such simple practices and skills as the use of improved varieties, improved planting practices, correct fertilizer application, etc.

Conclusion

At the end of Chapter 7, three main hypotheses were listed based on the information and data juxtaposition of two study institutions:

1. The roles of university for rural development (why a university needs to serve rural communities);
2. Effective model for transforming knowledge, technology, information, and skills from university to rural communities (what activities should be used by universities to carry out rural development programs);
3. Innovative approaches undertaken by university for rural development (how are these models realized in practice).

These three main hypotheses have several subhypotheses that can be identified under an "IF–Then" design. This means that if the "IF" conditions hold then the "THEN" conditions could follow. Hence, the various propositions, hypotheses, generalization can be analyzed under this arrangement. Thus in example I (hypotheses I), if three subhypotheses hold, then three could follow.

Those hypotheses have been repeated below in order to come to the conclusion.

Hypotheses

1. The roles of university for rural development (why a university needs to serve rural communities)

If

     i. Universities, especially agricultural universities in developing countries have clearly identified that rural development is their main mission. Universities transform their knowledge base from research and apply it into rural areas.

    ii. Different levels of government have committed strongly to support universities to deliver rural development service both financially, institutionally and with relevant strategies and policies.

     iii. Communities have paid great attention to the programs carried out by university for rural, remote and Indigenous areas development.

Then

     i. University has a key role to play in rural development.

     ii. University cannot play such a role if there is no policy support or financial aid from governments.

     iii. A university's rural development program cannot achieve the expected outcomes, if it fails to work with other institutions concerned to form a network serving rural development.

2. Effective model for transforming knowledge, technology, information, and skills from university to rural communities (what activities should be used by universities to carry out rural development programs).

If

     i. Establishing demonstration communities.

     ii. University's professors and staff are willing to go out of the campus and spend time and live in rural community.

     iii. University has set up a network and build up a partnership with relevant institutions and organizations to share resources so as to transform knowledge, technologies, and skills into rural communities.

     iv. University's training, research, and extension programs have been closely linked with the local needs and university has considered any specific conditions and situation in the program target areas.

     v. Apart from the university's contribution for rural development, the university itself has also grown while it serves rural development.

Then

     i. The university's rural development program can be more successful, effective, active, and efficient.

   ii. Local community members are more interested in participating in the program.

   iii. Expected outcomes can be reached.

   iii. Innovative approaches undertaken by university for rural development (how are these models realized in practice).

If

   i. Digital technology, Internet access, and other simple and effective media have been used by the university for its rural education and agricultural extension. Digital infrastructure has been extended from urban to rural areas. Then efficiency in terms of cost, staff's time, and learner's achievement will be much more increased.

   ii. Organizing community members into various technical or learning societies, associations, or other NGOs under the guidance of university staff. Then a learning society could be created.

   iii. The university–community partnership has been established and using package contract approach and establishing joint ventures.

   iv. Encouraging university student volunteers, especially agricultural university students, to launch social practice work and other practical courses in rural areas.

   v. Setting up rural, regional, or night training and consulting centers.

   vi. Training and encouraging a large group of community members to become backbone members of extension work force.

Then

   i. New findings and new skills according to practice needs can be put into the communities for their development.

   ii. Ensuring that all efforts are applicable, appropriate, and necessary for rural communities.

   iii. Community members' ability and capacity can be empowered as well as a learning society can be created so that long-term benefits can be achieved.

    iv. The rural development projects can be more sustainable when the project implementaters leave the project sites.

    v. Apart from serving the rural development, university itself can also be developed; university staff and students can learn from farmers get benefits from implementing the programs.

There is no doubt that universities, especially agriculture-oriented universities, can make and do positive and active contributions for rural development. It was also noted that universities themselves cannot fulfill this comprehensive and complex tasks alone due to some weaknesses that exist administratively, institutionally, financially, and with lack of government resources. Therefore, apart from universities adjusting their missions and concentrating on rural service and rural development through teaching, research, and extension/consultancy programs, governments at different levels should also rethink their own objectives and working areas to support the university's rural development programs. Furthermore, communities must have a positive attitude to co-operation with the university rural development programs and make the best use of them to serve the communities. Apart from those efforts, modern information technology, and digital media give more opportunities to enhance the university's rural development services.

Appendix 1: Interview Schedule

There are some interviews during the book writing, and the interviews are referred to in the text by their appropriate number. For example, Interview 2 will be referred to as In 2.

Code	Name	Position	Time	Place
In1	Yu Zongzhou	Professor of AUH	August 2001	Baoding, China
In 2	Sun Jianshe	Professor of AUH	August 2001	Baoding, China
In 3	Gu Ziling	Professor of AUH	September 2001	Baoding, China
In 4	Xue Qinglin	Professor of AUH	September 2001	Baoding, China
In 5	Zhang Pu	Professor of AUH	October 2001	Baoding, China
In 6	Wang Huijun	Professor of AUH	October 2001	Baoding, China
In 7	Yu Fuzeng	Former Secretary General, Chinese National Commission for UNESCO	December 2001	Beijing, China
In 8	Han Qinglin	Hebei Provincial Educational Bureau	December 2001	Shijiazhuang, China
In 9	Yu Zongzhou (re-interview)	Professor of AUH	February 2002	Baoding, China
In 10	Sun Jianshe (re-interview)	Professor of AUH	January 2002	Baoding, China

(Appendix Table Contd)

(Appendix Table Contd)

Code	Name	Position	Time	Place
In 11	Gu Ziling (re-interview)	Professor of AUH	January 2002	Baoding, China
In 12	Xue Qinglin (re-interview)	Professor of AUH	February 2002	Baoding, China
In 13	Zhang Pu (re-interview)	Professor of AUH	February 2002	Baoding, China
In 14	Wang Huijun (re-interview)	Professor of AUH	Feb. 2002	Baoding, China
In 15	Mr Antoine Barnaart	Pro-Vice Chancellor TAFE	June 2003	Darwin, Australia
In 16	Dr Greg Hill	Dean, Faculty of Education, Health and Science, CDU	June 2003	Darwin, Australia
In 17	Dr Allan Arnott	Senior lecturer of CDU	June 2003	Darwin, Australia
In 18	Dr Ian Falk	Professor of CDU	June 2003	Darwin, Australia
In 19	Dr Brian Devlin	Associate Professor of CDU	February 2004	Darwin, Australia
In 20	Dr Suzanne Parry	Senior lecturer of CDU	February 2004	Darwin, Australia
In 21	Dr Greg Hill (re-interview)	Dean, Faculty of Education, Health and Science, CDU	March 2004	Darwin, Australia
In 22	Dr Peter Wignell	Senior Lecturer, CDU	September 2004	Darwin, Australia
In 23	Dr Greg Shaw	Senior Lecturer, CDU	September 2004	Darwin, Australia
In 24	Mr Prue King and Ms. Jaclyn Miller	Remote Area Librarian and Head Librarian, Batchelor Institute	September 2004	Batchelor, Australia
In 25		NT government	September 2004	Darwin, Australia

Appendix 2: Survey Schedule

There are some surveys during the book writing, and the surveys are referred to in the text by their appropriate number. For example, Survey 2 will be referred to as SS 2.

Code	Name	Position	Time	Notes
SS 1	Greg Shaw	Senior Lecturer, CDU	September 2002	By email, English
SS 2	Yu Fuzeng	Former Secretary General, Chinese National Commission for UNESCO	September 2002	By email, English
SS 3	Glyn Rimmington	Professor, Uni of Wichita, USA	September 2002	By email, English
SS 4	Peter Jolly	NT government	March 2003	By email, English
SS 5	Jane Zhang	Lecturer, Batchelor Institute	March 2003	By email, English
SS 6	Ray Cleary	Associate Professor, Uni. of Wollongong	Jan. 2003	By email, English
SS 7	Helen Genandoy	PhD student, AUH	September 2002	By email, English
SS 8		Staff, CDU	September 2002	By email, English
SS 9		Staff, CDU	September 2002	By email, English

(Appendix Table Contd)

(Appendix Table Contd)

Code	Name	Position	Time	Notes
SS10	Zhang Tiedao	Beijing Academy of Educational Research, Beijing	September 2001	By email, Chinese
SS11	Wang Qiang	Nanjing Normal Uni. Nanjing,	September 2002	By email, Chinese
SS12	Tao Peijun	Staff, AUH	September 2002	By email, Chinese
SS13	Zhou Jizhu	Staff, AUH	September 2002	By mail, Chinese
SS14	Zhang Zhihua	Staff, AUH	March 2003	By email, Chinese
SS15	Zhou Damai	Staff, AUH	March 2003	By email, Chinese
SS16	Zhang Jianguang	Staff, AUH	July 2003	By email, Chinese

Appendix 3: Field Visit and Investigation Schedule

There are some field visits and investigations during the book writing, and the field visits and investigations are referred to in the text by their appropriate number. For example, Field Visit and Investigation 2 will be referred to as FI 2.

Code	Place	Time	Note
FI1	Baoding and ChaiChang, Hebei China	September 2001	Spend 10 days
FI 2	Qiannanyu, Hebei China	September 2001	Spend one week
FI 3	Anguo, Hebei China	January 2002	Spend two weeks
FI 4	Shijiazhuang, Hebei China	January 2002	Spend two weeks
FI 5	Baoding, Hebei China	December 2002	Spend one month
FI 6	Beijing, China	December 2003	Spend ten days
FI 7	Baoding, Hebei China	Jan 2004	Spend one month
FI 8	Baoding, Hebei China	June 2004	Spend one month
FI 9	Katherin, NT, Australia	November 2001	Spend one day
FI 10	Darwin, Australia	June 2001–August 2001	During my staying in CDU
FI 11	Darwin, Australia	November 2001–January 2002	During my staying in CDU
FI 11	Darwin, Australia	March 2002–December 2002	During my staying in CDU
FI 12	Darwin, Australia	February 2003–November 2003	During my staying in CDU
FI 13	Darwin, Australia	February 2003–June 2003	During my staying in CDU

Bibliography

Books and Periodicals

AAACE. Never to late to learn. A Report on Older People and Lifelong Learning. New South Wales Committee on Ageing. Australian Association of Adult and Community Learning (AAACE), 1997.

Adhikarya, R. "Strategic Extension Campaign: Increasing Cost-Effectiveness and Farmers' Participation in Applying Agricultural Technologies," 1995. www.fao.org/sd/EXdirect/EXan0003.htm (accessed January 10, 2002).

Ahmed, M., et al. "Education and Training for Rule Transformation: Skills, jobs, food and green future to combat poverty." Beijing, UNESCO-INRULED, 2012.

Agogino, Alice M. and Sherry Hsi."Learning Style Based Innovations to Improve Retention of Female Engineering Students in the Synthesis Coalition." Paper presented at ASEE/IEEE frontiers in Education '95 proceedings, Purdue University, 1995. http://fairway.ecn.purdue.edu/asee/fie95/4a2/4a21/4a21.htm (accessed in 2004).

Anderson, Jonathan. *Technology and Adult Literacy*. London, New York: Routledge, 1991.

Anderson, C. A. *Methodology of Comparative Education*. London: Macmillan, 1969.

Atchoarena, David and Lavinia Gasperini. "Education for Rural Development: Towards New Policy Responses." FAO and UNESCO. 2003.

AUH. "Technical Literate in Rural Area." Agricultural University of Hebei, 1995.

———. "The Collection of Reports on Taihang Mountain Road." Vol. 1, 1996.

———. "The Collection of Reports on Taihang Mountain Road." Vol. 2, 1997.

———. "The Research Team Report of 'Integration of the Three', Facing to Economy Construction, Carrying out the Integration of Teaching, Scientific Research and Production and the Direction of Education Reform," 1998.

Aulich, T. C. Come in Cinderella: The Emergence of Adult and Community Education. Report of the Senate Standing Committee on Employment, Education and Training. Senate Publications Unit, Parliament House, Canberra, 1991.

Bao, Jihong. Brief Introduction to Eradicating Illiteracy among Youths and Adults in Hebei Province. Department of Adult Education of Hebei Education Commission, 1997.

Bawden, B. A Learning Approach to Sustainable Agriculture and Rural Development: Reflections from Hawkesbury, 1996. www.fao.org/EXdirect/EXan0010.htm (accessed January 10, 2002).

Bell, J. *Doing Your Research Project*. Open University Press, Berkshire, England, 1993.

Bereday, G. Z. F. 'Reflection on Comparative Methodology in Education, 1964–1966," *Comparative Education*, 3, no. 3 (June 1967): 169–87.

———. *Comparative Method in Education*. USA: Holt, Rinehart and Winston, Inc, 1964.

Berzins, B. and Loveday. *The Northern Territory University and Preceding Institutions 1949–1999*. Northern Territory University Press, Darwin, Australia, 1999.

Billett, S. Ontogeny and Participation in Communities of Practices: A Socio-Cognitive View of Adult Development. *Studies in the Education of Adult*, 30. no. 1 (1998): 21–34.

BIITE. Batchelor Institute of Indigenous Tertiary Education. Annual Report, Batchelor, NT, Australia, 2001.

Batchelor. Handbook, Batchelor, NT, Australia, 2002.

Bird, K., D. Hulme, K Moore, and A. Shepherd. "Chronic poverty and remote rural areas." Chronic Poverty Research centre (CPRC) Working Paper no. 13, 2011.

Brundage , D. and Mackerarcher, D. (1980) *Adult Learning Principles and Their Application to program Planning*. Toronto: The Minister of Education.

Birkey, Richard C. and Joseph J. Rodman. Adult Learning Styles and Preference for Technology Programs, 1995. http://www2.nu.edu/nuri/llconf/conf1995/birkey.html (accessed on May 18, 2004).

Bogdan, R. and S. Biklen. "Qualitative Research for Education." Boston: Allyn and Bacon, 1982.

Bouma, G. *The Research Process*. Melbourne: Oxford University Press, 1993.

Brew, Cynthia A. "Teacher Education in England and N.S.W.: A Comparative Study of Problems of Transition from Student to Teacher," Master Thesis, University of Wollongong, Australia, 1980.

Brookfield, S. D. *Understanding and Facilitating Adult Learning* (p. vii). San Francisco, CA: Jossey-Bass, 1986.

Burns, R. B. *Introduction to Research Methods*. Melbourne: Longman, 1994.

Burrows, J. "University Adult Education in London." University of London, Senate House, 1976.

Cabanatan, Priscilla. "Contribution of Universities to Rural Agricultural Development through Science and Technology, Proceedings of UNESCO Asia and the Pacific Regional Meeting on the Role of Universities for Rural Development." INRULED, Baoding, China, 1998.

CAEFS. "Certificate in Access to Employment and Further Study." Northern Territory Department of Education, NT, Australia, 1991.

Canadian International Development Agency. http://acdi-cida.gc.ca/acdi-cida/ACDI-CIDA.nsf/eng/NAT-824104736-KCT (accessed on December 29, 2011).

Cantor, Jeffrey A. *Delivering Instruction to Adult Learners*. Toronto: Wall & Emerson, 1992.

Cavanagh, D. M. and G. W. Rodwell. *Dialogues in Educational Research*. NTU Printing/Publishing Services, Darwin, Australia, 1992.

Cavanagh, D. M., L. Connell, and M. Marriner. "Capacity Building in Rural Communities." Policy Concept Paper commissioned by the Chinese National Commission for UNESCO, and UNESCO-INRULED. Baoding, China, 2005.

Cavanagh, D. M., Lorraine Connell, Maria Marirner. "Professional Development and Training of the Teacher: Capacity Building Policy in Rural Communities." Presented at the proceedings of Rural Education for All in Sustainable Rural Development: Policy Analyses and Case Studies International Conference. International Research and Training Centre for Rural Education, UNESCO, Hebei, China, 2007.

Cavanagh, D. M. and Maria Marriner. "Theory Meets Practice in Rural Education and Training. Australian paper for UNESCO International Experts Seminar on Rural Education for Sustainable Development." Beijing, China, 2005.

Cavanagh, T. J and D. M. Cavanagh. "Interview and Power Point Presentation with a Business Entrepreneur: Seminar on Technical and Further Education." Chongqing, China, 2009.

Chambers, R. *Rural Development: Putting the Last First*. Brighton, UK: Longman Group Limited, 1983.

Chen, Jiyuan. *The Rural Socio-Economic Change in China*, 1st edition. Shanxi Economy Press.

Chen, Jinzhan and Paolian Cai. *Lifelong Education Theory and Adult Education Practice*. Capital Normal University Press, 1999 (original in Chinese).

Chen, Z., Z. Ren, et al. "Research on Essence and Feature of China's Economic Development Miracle Based on the Analysis of Route Evolution since Reform and Opening-up [J]." *Journal of Finance and Economics* 5, 2009.

Chinapah. V. "Education for Rural Transformation (ERT)—National, International, and Comparative Perspectives." Institute of International Education, Stockholm University, Universitets service US AB, Stockholm, 2011, pp. 33–47.

Chinapah, V. and L. Wang. Education for Rural Transformation through Principals' Professional Evelopment. Strategies to Achieve Balanced Inclusive Educational Development: Equity-Quality-Internationalization. V. Chinapah and L. Wang. Stockholm and Beijing, Universitetservice, Stockholm-Sweden, and UNESCO-INRULED, 2012, pp. 159–80.

Chinese National Bureau of Statistics. *Chinese National Statistic Handbook*, 2001.

Chow, P. C. Y. *China as the World Market and/or the World Factory in the Global Economy. China and the World Economy: China's Economic Rise After Three Decades of Reform*, Cambridge Scholars Publishing, 2011, 37–39.

Clark, M. C. and R. S. Caffarella. *An Update on Adult Development Theory: New Ways of Thinking about the Life Course. New Directions for Adult and Continuing Education,* no. 84 San Francisco: Jossey-Bass, 1999.

Cleary, R.P. "The Management of Rural Adult Education Agencies during Harsh Economic Times." Ph.D. thesis, University of Georgia, 1998.

Coles, Edwin K. Townsend. *Adult Education in Developing Countries.* Oxford, UK: Pergamon Press Ltd, 1977.

Dondas, M. *Profile Australia's Northern Territory,* 6th ed. Darwin, Northern Territory: Australia Lifestyle Publishing, 1998.

Cooparat, Pracob. "Information Technology as Strategies for Rural Development." Paper presented at the proceedings of UNESCO Asia and the Pacific Regional Meeting on the Role of Universities for Rural Development, INRULED, Baoding China, 1998.

Crowder, L. V. Agricultural Extension for Sustainable Development, 1998. http://www.fao.org/sd/EXdirect/EXan0004.htm (access dateJanuary 10, 2002).

Daloz, L. A. *Mentor: Guiding the Journey of Adult Learners.* San Francisco: Jossey-Bass, 1999.

De Vos, A. S., C. Delport, C. B. Fouche, and H. Strydom. *Research at Grass Roots: A Primer for the Social Science and Human Professions.* Pretoria: Van Schaik Publishers, 2011.

Department of Education Training and Youth Affairs. Innovative practices: Research and research training. *Higher Education Report for the 2000–2002 Triennium,* p. 150, 2000.

Delors, Jacques. "Learning: The Treasure Within, Report to UNESCO of the International Commission on Education for the Twenty-first Century." UNESCO, Paris, 1996.

Denzin, N. K. *Handbook of Qualitative Research.* Thousand Oaks, CA: SAGE Publications, 1994.

Deshler, D. "Evaluating Extension Programmes," Part 1, 1997. http://www.fao.org/sd/EXdirect/EXan0029.htm (accessed on January 10, 2002.

Dewar, Tammy. Adult Learning Online, 1996. http://www.cybercorp.net/~tammy/lo/oned2.html (accessed on May 16, 2004).

Dick, Gordon. "The Development of Katherine Rural College's Role in Severing the Needs of the Northern Territory." Northern Territory Department of Education, 1986.

Dirkx, J. "Transformative Learning Theory in the Practice of Adult Education: An Overview." *PAACE Journal of Lifelong Learning* 7 (1998): 1–14.

Donna, Williams-Sowter. "In Search of Best Practice: Learning Resources for Indigenous Tertiary Education—A Case Study." Batchelor Institute of Indigenous Tertiary Education, Batchelor, NT, Australia, 2002.

Donald, B.W. *Agricultural Extension.* Melbourne, Australia: Melbourne University Press, 1968.

Evenson, R. The Economic Contributions of Agricultural Extension to Agricultural and Rural Development, 1997. http://www.fao.org/sd/EXdirect/EXan0038.htm (accessed on January 10, 2002).

FAO. "Integrating Agricultural Research, Education and Extension in Developing Countries," 1996a. www.fao.org/sd/EXdirect/EXan0009.htm (accessed on January 10, 2002).

————. "Partners in Sustainable Development: Linking Agricultural Education Institutions and Farmer Organizations," 1996b. http://www.fao.org/sd/EXdirect/EXan0007.htm (accessed on January 10, 2002).

————. Assessment of Pre-Service Extension Education, 1996c. http://www.fao.org/sd/EXdirect/EXan0001.htm (accessed on January 10, 2002).

————. The role of non-governmental organizations in extension. In B.E. Swanson, R.P. Benz, and A.J. Sofranko (eds), *Improving Agricultural Extension: A Reference Manual*, 3rd ed. Rome: FAO (Food and Agricultural Organization of the United Nations), 1997.

Farrington, J. "The Role of Non-Government Organizations in Extension, 1997. http://www.fao.org/sd/EXdirect/EXan0040.htm (accessed on January 10, 2002).

Feagin, J., A. Orum, and G. Sjoberg. *A Case for Case Study*. Chapel Hill. NC: University of North Carolina Press, 1991.

Feng, Zengzun. *Comparative Education*. Jiangsu Educational Press, Jiangsu, PRC, 2001.

Gammon, E. The History and Development of Government School in the Darwin Region, 1992.

Gasperini, L. "From Agricultural Education to Education for Rural Development and Food Security: All for Education and Food for All", 2000. http://www.fao.org/sd/EXdirect/EXre0028.htm (accessed onJanuary 10, 2009).

Genandoy, Helen and Fuzeng Yu. "The Contribution of Higher Education in the Development of Rural Areas." UNESCO-INRULED, Baoding, China, 2001.

Griffith, D. "Quantifying Access to Services in Remote and Rural Australia in Education, Equity and the Crisis in the Rural Community." Paper presented at the proceedings of the Rural Education Research Association Conference, Alie Springs, 1992.

Guan, Chunyun. "Hunan Agricultural University to Serve Local Economic Development." Paper presented at the proceedings of UNESCO Asia and the Pacific Regional Meeting on the Role of Universities for Rural Development, INRULED, Baoding China, 1998.

Guo, Shutian. *Food Policy: Theory and Positivism*, 1st edition. Xinhua Press, Xinhua, PRC, 1995.

Harbison, F. H. "The Development of Human Resources. An Analytical Outline." *Economic Development in Africa*. Edited by E. F. Jackson. Basil Blackwell, Oxford, UK, 1965.

Harrison, B. "Final Count of Persons by Age and Sex for Statistical Local Areas: Northern Territory." Australia Bureau of Statistics, Darwin, N.T, 1992.

Hartman, Virginia F "Teaching and Learning Style Preferences: Transitions through Technology." *VCCA Journal* 9, no. 2 (Summer 1995). http://www.so.cc.va.us/vcca/hartl.htm (accessed on May 18, 2004).

Hobson, P. Adult Development and Transformative Learning. *International Journal of Lifelong Education.* Vol. 17, No. 2, pp. 72–76, 1998.

Holster, J. The Education of Adult, Studies in Adult Education, 9, no. 1 (1977): 52–61.

IESP. "Indigenous Education Strategic Plan 2000–2004." Northern Territory Department of Employment, Education and Training, NT, Australia.

ILO. Decent work agenda, 2012. http://www.ilo.org/global/about-the-ilo/decent-work-agenda/lang—en/index.htm (acessed on December 28, 2011).

INRULED. *Higher Education for the Development of Mountainous Areas.* Baoding, China: INRULED publication, 2000.

———. *Education for Rural Transformation: Toward a Policy Framework.* Baoding, China: INRULED Publication, 2001.

Jarvis, P. *Adult and Continuing Education,* Breckenham, Kent: Croom Helm, 1983.

Karmel, P. K. P. "Public Policy and Higher Education Public Policy and Higher Education," *Australian Journal of Management* 26 (August 2001): 124.

Karyadi, Darwin. "Poverty Alleviation Strategies: Indonesian Case." Paper presented at the proceedings of UNESCO Asia and the Pacific Regional Meeting on the Role of Universities for Rural Development, INRULED, Baoding China, 1995.

Khan, Yar Muhammad "Role of University in the Rural Development, in the Context of 21 Century." Paper presented at the proceedings of UNESCO Asia and the Pacific Regional Meeting on the Role of Universities for Rural Development, INRULED, Baoding, China. 1998.

Kilpatrick, S., J. Abbott-Chapman et al. "Identifying the Characteristics of Rural Learning Communities: Implications for Rural Development." SPERA, Society for the Provision of Education in Rural Australia, 2011.

Knowles M. *Informal Adult Education.* Chicago: Association Press, 1950.

———. *The Adult Learner: A Neglected Species,* 4th edition. Houston, TX: Gulf Publishing, 1990.

Kramer-Koehler, Pamela, Nancy M. Tooney and Devendra P. Beke."The Use of Learning Style Innovations to Improve Retention." Paper presented at the proceedings of ASEE/EEE Frontiers in Education '95. Purdue University. http://fairway.ecn.purdue.edu/asee/fie95/4a2/4a22/4a22.htm (accessed on May 18, 2004).

Kuchinke, K. P. "Adult Development towards What End? A Philosophical Analysis of the Concept as Reflected in the Research, Theory, and Practice of Human Resources Development," *Adult Education Quarterly* 49, no. 4 (Summer 1999): 148–62.

LABORSTA, Labour Statistics Database, International Labour Organization (ILO), Geneva. http://laborsta.ilo.org/ (accessed on December 28, 2011).

Li, Shaoyuan. *Rural Education*. China: Jiangsu Educational Press, 2000.

Li, Xiaoyun. "What Could the Universities Do Directly for Rural Development? The Lessons and Perspectives." Presented at the proceedings of UNESCO Asia and the Pacific Regional Meeting on the Role of Universities for Rural Development, INRULED, Baoding China, 1998.

Loveday, P. and D. Wade-Marshall. "Economy and People in the North." Australia National University, 1985.

Lu, Yuefeng. "Mechanism of the Integration of Agriculture, Science and Education, Educational Study," *Educational Research*, Vol. 17, no. 10, pp. 24–40, 1996 (original in Chinese).

Luo, Xiwen. "The Cooperation of Agricultural University with Farmer's Enterprise for the Promotion of Agricultural Industrialization." Presented at the proceedings of UNESCO Asia and the Pacific Regional Meeting on the Role of Universities for Rural Development, INRULED, Baoding, China, 1998.

Maguire, C. J. From Agriculture to Rural Development: Critical Choices for Agriculture Education, 2000. http://www.fao.org/sd/EXdirect/EXre0029.htm (accessed on May 26, 2004).

Majumdar, S. "Emerging Trends in TVET in Asia and the Pacific Region: CPSC, Åôs Response." *Emerging Challenges and Trends in TVET in the Asia-Pacific Region*, UNESCO, 2011, 3–17.

Merriam, S. B. and R. S. Caffarella. *Learning in Adulthood*, San Francisco: Jossey-Bass, 1999.

Merriam, S. B. and E. L.Simpson. *A Guide to Research and Educators and Trainers of Adults*. Malabar, FL: Krieger Publishing Company, 1989.

Muny, P. "Trends and Needs in Manpower Planning for Sustainable Agricultural and Rural Development in Cambodia: An Educational Planning Assessment," 1997. www.fao.org/sd/EXdirect/EXre0017.htm (accessed on January 10, 2002).

Ning, An. "The Challenge and Expectations of China's Agricultural Education and Development." Paper presented at the proceedings of UNESCO Asia and the Pacific Regional Meeting on the Role of Universities for Rural Development, INRULED, Baoding China, 1998.

Noguchi, Noboru. "Opening Speech." Presented at the proceedings of UNESCO Asia and the Pacific Regional Meeting on the Role of Universities for Rural Development, INRULED, Baoding, China, 1998.

Nyerere, J. K. "Education for Self-reliance." *Adult Education Handbook*. Tanzania Publishing House, Tanzania, Africa, 1973.

Oakley, P. and C. Garforth. "Guide to Extension Training, Food and Agriculture Organization of the United Nations." Rome, 1985.

Ordonez, Victor, P. K. Kasaju, and C. Seshadri. "Basic Education for Empowerment of the Poor." UNESCO Principal Regional Office for Asia and the Pacific, Bangkok, Thailand, Proap, Bangkok, 1998.

Patton, M. Q. *Qualitative Evaluation Methods*. Beverly Hills, CA: SAGE Publications, 1980.

———. *Qualitative Evaluation and Research Methods, CA*. London: SAGE, 1990.

Ping, C. J. Graduate Perspectives in a Changing Society, UNESCO, Paris, 1998.

Rahman, Mohammad H. "The Case of Bangladesh." Paper presented at the proceedings of UNESCO Asia and the Pacific Regional Meeting on the Role of Universities for Rural Development, INRULED, Baoding, China, 1998.

Raman, K. V. "Role of Universities: Indian Scenario." Paper presented at the proceedings of UNESCO Asia and the Pacific Regional Meeting on the Role of Universities for Rural Development, INRULED, Baoding, China, 1998.

Reading, C. et al. "Focusing on ICT in Rural and Regional Education in Australia." Australian Educational Computing, 2006.

Sakya, T. M. "Post-Literacy Programmes," UNESCO Principal Regional Office for Asia and the Pacific Bangkok, Thailand, 1993.

Sen, Amartya. Poverty and Famines: An Essay on Entitlement and Deprivation. Oxford University Press, Oxford University Press, UK, 1981.

Shaw, Greg. "The Adult Learning and Experience: Learning in Mixed-Mode Forms of Professional Development in the Northern Territory of Australia." Ph.D. thesis in the Faculty of Education, Deakin University, Victoria, Australia, 1999.

Singh, Madhu. "Adult Learning and the Changing World of Work," UNESCO Institute of Education, 1998. www.unesco.org/education/uie/pdf/madhu1.pdf (accessed on May 18, 2004).

Sinlarat, Paitoon. "Role of the Thai Universities in Rural Development: Time for New Concepts and Methods," Paper presented at the proceedings of UNESCO Asia and the Pacific Regional Meeting on the Role of Universities for Rural Development, INRULED, Baoding, China, 1998.

Smith, R.M. *Learning How to Learn*. Chicago, IL: Follett Publishing Company, 1982.

Sommer, K. N. "Education and Food for All," 2001. http://www.fao.org/food/tf2000/armen-e.htm (accessed on May 26, 2003).

Strauss, A. and J. Corbin. *Basics of Qualitative Research-Grounded Theory Procedures and Techniques*. CA: SAGE, 1990.

TAFETPS. "Technical and Further Education Triennial Planning Submission for 1985–87." Northern Territory Department of Education, NT, Australia, 1983.

Takwale, Ram. "Experience of and Expectations from Indian Universities." Paper presented at the proceedings of UNESCO Asia and the Pacific Regional Meeting on the Role of Universities for Rural Development, INRULED, Baoding, China, 1998.

Tan, Kim, et al. "The University and Vocational Training in Rural Areas: A Case Study." Paper presented at the proceedings of the International Workshop on Technical Training for Rural Development toward the 21st Century, China Agricultural Science and Technology Press, 2001, 252–57.

Teh, Wan Hashim Wan. "Role of Universities for Rural Development: The Malaysia Experience." Paper presented at the proceedings of UNESCO Asia and the Pacific Regional Meeting on the Role of Universities for Rural Development, INRULED, Baoding, China, 1998.

Tellis, Winston. "Application of a Case Study Methodology." *The Qualitative Report*, 3, no. 3 (September 1997). http://www.nova.edu/ssss/QR/QR3-3/tellis2.htm (accessed on May 26, 2004).

Trochim, W. M. K. "Research Methods Knowledge Base." Cornell University, Ithaca, New York.

Tuckman, B. W. "Conducting Educational Research." Florida State University.

UNDP. "Human Development Report." UNDP, 1991.

UNESCO [PROAP]. "Higher Education in Transition Economies in Asia," 1998.

UNESCO Principal Regional Office for Asia and the Pacific. *Effective Implementation of Continuing Education at the Grassroots.*" Printed in Thailand, 52–4, 2001.

Yin, R. *Case Study Research: Design and Methods*. Thousand Oaks, CA: SAGE Publishing, 1994.

Ying, Wenyong. *Vocational and Technical Education and Socio-economic Development*. Yunnan Education Press, Yunnan, PRC, 1993.

Wang Li, Zhao Zhiqiang and Wang Lifang. "To Promote Poverty Alleviation through Technique Extension and Adult Education." INRULED, 1998.

———. "Income Generating Programs (IGP) for Rural Development in Hebei Province." INRULED Publication, 1999.

———. "Rural Technological Literacy: A Key to Promote Regional Economic Development." INRULED, 2000.

Wang, Xiaobing et al. "College Education and the Poor in China: Documenting the Hurdles to Education Attainment and College Matriculation." REAP, 2011.

Wellington, Jerry. *Educational Research*. London and New York: Continuum, 2000.

White, E. and J. Brands. "An Ecology of Relationship: Language, Understanding and Education." Batchelor Institute of Indigenous Tertiary Education, Batchelor, NT, Australia, 1999.

Wiersma, William. "Research Methods in Education." University of Toledo, Allyn and Bacon, 2000.

Wyn, Johanna, Ian Falkand John Guenther. Education for Rural Development in Australia 1945–2001, UNESCO International Research and Training Centre for Rural Education, 2002.

Yao, Jinguan and others. *Agricultural Product Circulating System and Pricing System in China*, 1st edition. China Price Press, 1993.

Zhang, Liuzheng. *The Future Structure of Chinese Economy*, 1st edition. PRC: CITIC Press, Peking University, 1992.

Zhang, Pu and Runzhi Su. *The History of the Agricultural University of Hebei*. Beijing, PRC: Social Science Development Press, 1992.

Zhang, Tiedao. "Toward a Responsive Education System for Rural Development, Regional Workshop on the Urgent Education Needs for Rural Development." Baoding, Hebei, China, November 2–11, 1994.

Zhang, Tiedao. "Empowerment of Villages with Technologies." Presented at the proceedings of the International Workshop on Technical Training for Rural Development Towards the 21st Century." Baoding, China, September 17–21, 2001.

Zhou, Zhihua, Chun Li and Pu Zhang. *Taihang Mountain Development Road*. Baoding, China: Hebei University Press, 1990.

Zinnah, M. M. "From Margin to Mainstream: Revitalization of Agricultural Extension Curricula in Universities and Colleges in Sub-Saharan Africa." 1998 http://www.fao.org/sd/EXdirect/EXan0027.htm (accessed on January 10, 2002).

Zou, Xinliang. "Rural Vocational Education and Socialist Market Economy." Journal of Fuzhou Teachers' College, no. 2(1996).

Websites

http://www.cdu.edu.au/visiting/abouthistory.html (accessed onMay 10, 2010).

http://www.utoronto.ca/writing/litrev.html (accessed on May 26, 2010).

http://www.hebau.edu.cn/xuexiaogaikuang/nongdajianjei.html (accessed on March 16, 2004)

Website 5.6 www.nt.gov.au/ntt/economic (accessed on November 13, 2010).

http://www.batchelor.edu.au/file/webpage/about_history.html (accessed on July 8, 2010).

http://www.batchelor.edu.au/file/webpage/about_structure.html (accessed on July 8, 2010).

http://www.batchelor.edu.au/index.html?a=1 (accessed on July 8,2003).

http://www.cdu.edu.au/orgchart/orgchart.html (accessed on May 8 2003).

http://www.clickforaustralia.com/MapoofNorthernTerritory.htm, (accessed on July 8, 2010).

http://www.dotrs.gov.au/regional/northern_forum/formal_response/top_end/research.htm, (accessed on August 28, 2010).

Documents

AAACE. Australia Association of Adult and Community Education, 2002. http://www.aaace.org/ (accessed in 2004).

Adult Education in China. *Department of Educational Development Planning under the Ministry of Education: China Statistical Yearbook of Educational Funds (1998–2007)*, China Statistics Press, Beijing, 2001.

AUH Information Handbook. 2001.

Theoretical and Practical Research on "the Project of Revitalizing Villages through Science and Education". Chinese Agriculture Society, Agriculture Press of China, 1999.

Chinese Educational Technology. Department of International Cooperation and Exchanges, Ministry of Education, China, 2001.

China Food Development Strategic Measures, 1st edition, Agriculture Press.

China Higher Education Law, 1990. www.chinaedu.net (accessed on August 26, 2003).

Action Scheme for Invigorating Education Towards the 21st Century. Chinese Ministry of Education, 1998.

Chinese People's Daily Newspaper, October 5, 1998; October 12, 1998, October 10, 1998, October 5, 1998.

Song Wei, Chen Baiming, Zhang Ying (2014). Landuse change and socio economic driving forces of rural settlement in China, from 1996-2005. China Geographical Science, 24(5) 511–524.

Declaration of UNESCO Asia and Pacific Regional Conference on National Strategies and Regional Co-operation for the 21st Century. Tokyo, Japan, July 8–18, 1997.

Declaration about Higher Education in Asia and the Pacific, Higher education National Strategies and Regional Co-operation for the 21st century. Tokyo, Japan, July 8–10, 1997

DEET Annual Report 2001–2002, Department of Employment, Education and Training of Northern Territory, 2000, p. 115.

Education for All Action Plan of Hebei Province (1993–2000), Hebei Provincial Education Bureau.

Education Law of the People's Republic of China. State Education Ministry of the People's Republic of China Beijing, 1999.

Final Report of UNESCO Asia and the Pacific Regional Meeting on the Role of Universities for Rural Development. INRULED, Baoding China, 1998.

Final Report on Human Rights and Extreme Poverty. United Nations Economic and Social Council, 1996. http://www.unhchr.ch/huridocda/huridoca.nsf/ (Symbol)/E.CN.4.Sub.2.1996.13.En?Opendocument (accessed in 2004).

Forty Years of Hebei Countryside. Hebei People Press, 1st edition.

General Information of Batchelor Institute, 1990. http://www.batchelor.edu. au/public/document/HbkBatchelor2003/Intro/General_Information.pdf (accessed on July 8, 2003).

Handbook of Batchelor Institute. http://www.batchelor.edu.au/public/document/ HbkBatchelor2003/Intro/Introduction.pdf (July 8, 2003).

Hebei Agriculture Environment Quality Report. Hebei Agriculture Environment Protection Monitoring Station, Hebei Agriculture Ecology Society, 1996.

Higher Education Report for the 2001 to 2003 Triennium. Australia Department of Education, Training and Youth Affairs, Australia, p. 3, 2001.

Lifelong Education Theory and Adult Education Practice. Capital Normal University Press, 1999.

Northern Territory Department of Education Annual Report 1979/1980, NT, Australia. NTRC Pamphlet.

Poverty and Financial Hardship Report.Committee Hansard 4.8.03, p. 1193 (QCOSS). www.aph.gov.au/~/media/.../committee/clac.../poverty/.../c14_pdf. ashx

Poverty and Financial Hardship Report.Committee Hansard 2003, p. 1081 [NTCOS) www.aph.gov.au/~/media/wopapub/.../committee/clac.../c14_pdf. ashx

Poverty and Financial Hardship Report.Committee Hansard 28. 7.03, p. 1066 Han. Jonathan Ford MLC www.parliament.wa.gov.au › Home › Members

Poverty and Financial Hardship Report . Catholic Welfare Australia. Submission 148, p. 29. www.aph.gov.au/~/media/wopapub/senate/.../clac.../c14_pdf.ashx

Poverty and Financial Hardship Report . Catholic Welfare Australia. Submission 148, p. 30 www.aph.gov.au/~/media/wopapub/senate/.../clac.../c14_pdf.ashx

The World Food Summit. 1996 www.who.int/trade/glossary/story028/en/

The Laws on Education of the People's Republic of China, 1999, p. 92 http:// www.chinaeducenter.com/en/cedu/cel.php

Selected Northern Territory TAFE Statistics 1990, 1991, Northern Territory Department of Education, Darwin, NT, Australia.

Teacher Education in China. Department of International Cooperation and Exchanges, Ministry of Education, China, 2001.

Facing to Economy Construction, Carrying out the Integration of Teaching, Scientific Research and Production is the Direction of Education Reform. The Research Team of ' Integration of the Three", Agricultural University of Hebei, Baoding, Hebei, China, 1984.

Agricultural Resource Investigation Report of Taihang Mountain Areas. The Research Team on Resource Investigation of Taihang Mountain Areas of Agricultural University of Hebei, Baoding, Hebei, China, 1984.

University Information Handbook. Agricultural University of Hebei, Baoding, Hebei, China, 2001.

Vocational Education in China. Department of International Cooperation and Exchanges, Ministry of Education, China, 2001.

World Development Report, 2000/2001, Attacking Poverty, World Bank. http:// www.worldbank.org/poverty/wdrpoverty/repoh rt/index.tm (accessed in 2004).

World Bank Report, 2001, Reaching the Rural Poor, World Bank. http://web. worldbank.org/WBSITE/EXTERNAL/NEWS/0, contentMDK:20076031~ menuPK:34457~pagePK:34370~piPK:34424~theSitePK:4607,00.html (accessed in 2004).

Zemke, Ron and Susan Zemke. "30 Things We Know for Sure about Adult Learning," *Training* (July 1988): 57–61.

Index

About the Author

Li Wang is currently a Professor and Deputy Director at UNESCO International Research and Training Centre for Rural Education (INRULED) in China. Professor Wang's areas of interest include rural education and rural development, water resources, rural and town planning, and geothermal utilization. He has contributed to more than 60 books and papers, undertaken about 50 international research projects, and organized/participated in more than 100 international conferences, symposia, forums, and workshops. Professor Wang has represented INRULED and Chinese Ministry of Commerce in projects in Cambodia and Pakistan and in many international conferences, forums, and meetings.